Intelligent Automated Defect Detection

This book presents cutting-edge research on the integration of artificial intelligence, deep learning, and computer vision with traditional non-destructive testing (NDT) methods. It compiles recent advances in automated defect detection across critical infrastructure and industrial applications.

Intelligent Automated Defect Detection: AI Solutions for Roads, Welding, and Industrial Inspection demonstrates how AI-driven solutions are transforming quality inspection from subjective, time-intensive manual processes into real-time, scalable, and objective automated systems. The book provides comprehensive coverage of machine learning applications in civil infrastructure inspection and industrial welding quality assurance, utilizing radiography, ultrasonic B-scans, and time-series sensing. The authors present state-of-the-art deep learning architectures including convolutional neural networks, transformer models, and foundation models like the Segment Anything Model (SAM), alongside practical guidance on developing explainable AI systems and overcoming data scarcity challenges. The book offers proven methodologies for transitioning from proof-of-concept studies to robust real-world applications that enhance inspection efficiency while reducing costs and failure risks. Real-world case studies are included throughout, with additional applications available on the authors' GitHub page.

This book is intended for researchers, engineers, and practitioners working at the intersection of AI, NDT, and industrial quality assurance. It also serves as a resource for graduate students and early-career professionals interested in applying modern data-driven approaches to complex engineering problems.

Intelligent Automated Defect Detection

AI Solutions for Roads, Welding, and Industrial Inspection

Hassan Zargarzadeh
Amir-M. Naddaf-Sh
Sadra Naddaf-Sh
and
M-Mahdi Naddaf-Sh

CRC Press
Taylor & Francis Group
Boca Raton London New York

CRC Press is an imprint of the
Taylor & Francis Group, an **informa** business

Designed cover image: Shutterstock

First edition published 2027
by CRC Press
2385 NW Executive Center Drive, Suite 320, Boca Raton FL 33431

and by CRC Press
4 Park Square, Milton Park, Abingdon, Oxon, OX14 4RN

CRC Press is an imprint of Taylor & Francis Group, LLC

ISBN: 978-1-041-28797-1 (hbk)
ISBN: 978-1-041-28798-8 (pbk)
ISBN: 978-1-003-76657-5 (ebk)

DOI: 10.1201/9781003766575

Typeset in Nimbus Roman font
by KnowledgeWorks Global Ltd.

Publisher's note: This book has been prepared from camera-ready copy provided by the authors.

Dedication

The authors dedicate this book to their parents and family members, whose unwavering support, encouragement, and patience sustained them throughout the development of this work.

Contents

SECTION II Digital X-Radiography

SECTION III Arc Stud Welding

SECTION IV Ultrasonic Testing

Foreword

Infrastructure is the silent backbone of modern civilization. Roads carry commerce and connect communities. Welded structures sustain energy systems, transportation networks, and industrial production. Yet the integrity of these critical assets depends on our ability to detect, interpret, and respond to defects early, accurately, and at scale.

For decades, defect detection in civil and industrial systems relied heavily on manual inspection, heuristic image processing, and rule-based analysis. While effective in controlled settings, such methods struggle to meet the growing demands of modern infrastructure: real-time decision-making, large-scale data processing, cost efficiency, and safety. The convergence of artificial intelligence, deep learning, and advanced sensing technologies has fundamentally reshaped what is possible.

Intelligent Automated Defect Detection: AI Solutions for Roads, Welding, and Industrial Inspection arrives at a pivotal moment in this transformation. This book does more than apply deep learning models to defect detection. It builds an integrated, end-to-end engineering framework. Beginning with optimized convolutional neural networks for real-time road crack mapping, the authors move beyond simple binary classification and introduce geometrical mapping, hyperparameter optimization via the Bayesian methods, and deployable real-time systems. The progression into digital radiography, arc stud welding, ultrasonic B-scans, transformer-based detectors, and synthetic data generation demonstrates both breadth and technical rigor.

A defining strength of this work is its commitment to engineering realism. The methodologies presented are not abstract algorithmic exercises; they are grounded in:

- Carefully curated datasets
- Hardware-aware optimization
- Explainable AI techniques
- Real-time deployment constraints
- Industrial protocols and inspection standards

The inclusion of explainability in radiographic and welding applications reflects a mature understanding that industrial AI must not only perform accurately but also justify its decisions. The treatment of synthetic data, GANs, diffusion models, knowledge distillation, and transformer-based object detectors further places this book at the forefront of applied AI research.

Equally important is the interdisciplinary nature of the contributions. By bridging civil engineering, computer science, electrical engineering, non-destructive testing, and industrial manufacturing, the authors provide a unified perspective on defect intelligence across domains. The transition from pavement cracks to weld radiography and ultrasonic inspection illustrates a central thesis: **defect detection is**

fundamentally a pattern-recognition problem enhanced by domain-specific physics and engineering constraints.

This book will serve multiple audiences:

- Researchers seeking advanced AI methodologies for industrial inspection
- Graduate students entering the field of intelligent infrastructure systems
- Practicing engineers modernizing non-destructive testing workflows
- Industry leaders exploring real-time AI deployment in operational environments

Beyond its technical contributions, this book represents a shift in philosophy from post-failure analysis to predictive, automated, and scalable defect intelligence. As infrastructure systems age and industrial complexity increases, such tools will not merely enhance efficiency; they will become essential to safety, sustainability, and economic resilience.

The work presented here stands as a thoughtful and rigorous contribution to the emerging discipline of AI-driven inspection engineering. It reflects both scholarly depth and practical foresight, offering a roadmap for the next generation of intelligent monitoring systems. It is a pleasure to introduce this important contribution to the field.

Preface

The rapid development of artificial intelligence, computer vision, and advanced sensing technologies is transforming the way defects are detected, evaluated, and managed across civil infrastructure and industrial systems. Traditional inspection practices have long depended on trained personnel, manual interpretation, and rule-based processing methods. Although these approaches remain essential, they can be time-consuming, difficult to scale, and sensitive to differences in experience, working conditions, and human judgment. At the same time, aging infrastructure, increasingly complex manufacturing processes, and growing safety expectations have created a need for inspection systems that are faster, more consistent, and capable of processing large volumes of data.

This book was developed in response to that need. Its purpose is to bring together recent advances in artificial intelligence, deep learning, computer vision, and nondestructive testing within a unified engineering context. Rather than treating artificial intelligence as an isolated computational subject, the book examines how data-driven methods can be integrated with established inspection technologies to address practical defect-detection problems. The emphasis throughout is on methods that improve inspection accuracy, efficiency, scalability, and interpretability while remaining connected to the physical systems and sensing processes from which the data originate.

The applications presented in this book cover both civil infrastructure and industrial quality assurance. They include road-surface inspection, digital radiography, arc stud welding, and ultrasonic testing. These application areas differ substantially in their sensing methods, data formats, defect characteristics, and operational requirements. Road inspection commonly relies on images acquired from cameras in changing environmental conditions. Radiographic inspection involves the interpretation of internal structural features represented in X-ray images. Welding-quality assessment may require the analysis of images, time-series measurements, or combinations of process signals. Ultrasonic inspection presents additional challenges associated with B-scan interpretation, noise, limited labeled datasets, and variations in defect appearance. Despite these differences, the underlying objective is shared: to convert complex inspection data into accurate, timely, and meaningful information about the condition of a structure or component.

The book begins with an introduction to the role of intelligent automated defect detection and the broader relationship between artificial intelligence and engineering inspection. Section I focuses on roads. It presents optimized convolutional neural-network methods for real-time road-crack mapping and deep-learning approaches for efficient road-damage detection. These chapters address important practical considerations such as image acquisition, dataset preparation, model architecture, hyperparameter optimization, computational efficiency, and the transition from defect identification to useful spatial information.

Section II addresses digital X-radiography. The chapters in this section examine artificial intelligence–based weld assessment, deep-learning methods for radiographic defect analysis, and explainable real-time object detection in two-dimensional radiographic images. Radiographic inspection is a particularly important application because accurate interpretation requires both image-processing capability and an understanding of defect morphology. The methods presented illustrate how learning-based systems can support the identification and classification of weld defects while also providing information that helps users understand the basis for model decisions.

Section III focuses on arc stud welding. It explores defect classification using multivariate time-series measurements and machine-learning methods for automotive stud-weld quality assessment. These chapters demonstrate that automated inspection is not limited to image-based data. Process signals can also contain valuable information about weld formation, equipment behavior, and defect conditions. The section therefore considers signal preprocessing, feature representation, classification, model optimization, and explainability in the context of manufacturing-quality monitoring.

Section IV examines ultrasonic testing, with particular attention to industrial ultrasonic B-scan images. It presents deep-learning approaches for weld-defect detection, the use of the Segment Anything Model for defect segmentation, and synthetic-data-driven methods for defect classification. This section reflects several important directions in contemporary artificial intelligence, including transformer-based architectures, foundation models, parameter-efficient fine-tuning, generative modeling, and knowledge distillation. It also addresses one of the most persistent challenges in industrial inspection:

About the authors

Hassan Zargarzadeh is an Associate Professor in the Electrical and Computer Engineering Department at Lamar University, Beaumont, TX. He holds a PhD from Missouri University of Science and Technology (MS&T). His research focuses on applied machine learning in the fields of defect detection and robotics, where he has published several papers. He also has published numerous papers in the fields of advanced control systems and power electronics.

Amir-M. Naddaf-Sh is currently a Ph.D. student in Computer Science at Kennesaw State University, Marietta, GA. He received his B.Sc. degree in Electrical Engineering from Ferdowsi University of Mashhad, Mashhad, Iran, in 2020, and his Doctor of Engineering degree in Electrical Engineering from Lamar University, Beaumont, TX, USA, in 2025. His research interests span deep learning, robotics, and computer vision.

Sadra Naddaf-Sh is a Senior Machine Learning Engineer and Researcher at Truveta, Bellevue, WA, where he works on advancing efficient and robust deep learning and generative AI methods. His research interests include generative AI, language modeling, and data-centric AI. He received his Doctorate degree in Electrical Engineering from Lamar University, Beaumont, TX, USA, in 2022, and his B.Sc. degree in Computer Engineering from Ferdowsi University of Mashhad, Iran, in 2018.

M-Mahdi Naddaf-Sh is a Research Scientist at the Laboratory for AI, Robotics, and Automation (LARA) at the University of California, Davis, CA. His research spans the intersection of artificial intelligence, robotics, and control systems, with a focus on autonomous systems and machine learning for robotic perception and decision-making. Before joining UC Davis, he worked at Ford Greenfield Labs in Palo Alto, CA, where he led the Advanced Robotics and Automation Lab within the Core AI/ML group. He received his bachelor's degree in electrical engineering from Ferdowsi University of Mashhad (Iran), his Master's in Artificial Intelligence from the University of Southampton (UK), and his Doctor of Engineering in Electrical Engineering from Lamar University (USA). He has authored several publications in peer-reviewed IEEE journals and conferences, with contributions recognized in the areas of autonomous robotics and intelligent manufacturing.

Acknowledgments

SeyedSaeid Hosseini
Department of Civil Engineering, Lamar University
Beaumont, TX, USA

Jing Zhang
Department of Computer Science, Lamar University
Beaumont, TX, USA

Nicholas A. Brake
Department of Civil Engineering, Lamar University
Beaumont, TX, USA

Amir R Kashani
Artificial Intelligence Lab, Stanley Oil & Gas, Stanley Black & Decker
New Britain, CT, USA

Sayyed M Zahiri
Georgia Institute of Technology
Atlanta, Georgia, USA

Maxim Dalton
Artificial Intelligence Lab, Stanley Oil & Gas, Stanley Black & Decker
New Britain, CT, USA

Gabriel Elpers
Artificial Intelligence Lab, Stanley Oil & Gas, Stanley Black & Decker
New Britain, CT, USA

Soodabeh Ramezani
Artificial Intelligence Lab, Stanley Oil & Gas, Stanley Black & Decker
New Britain, CT, USA

Vinay S. Baburao
CRC-Evans
Houston, TX, USA

Zina Ben-Miled
Phillip M. Drayer Electrical and Computer Engineering Department, Lamar University
Beaumont, TX, USA

1 Introduction

Infrastructure and industrial assets form the backbone of modern civilization. Roads enable economic mobility, welded joints ensure structural integrity in pipelines and pressure vessels, and industrial components sustain energy production, manufacturing, and transportation systems. The reliability and safety of these systems depend heavily on early and accurate defect detection. Failures in pavement surfaces, weld seams, or structural components are not merely cosmetic deficiencies but they are precursors to financial loss, environmental damage, safety hazards, and, in extreme cases, catastrophic failure.

For decades, defect detection has relied primarily on manual inspection and traditional non-destructive testing (NDT) techniques. Although effective in many contexts, manual inspection is labor-intensive, time-consuming, subjective, and often inconsistent. In roadway surveys, inspectors visually identify cracks and surface distresses under varying environmental conditions. In welding and industrial inspection, experts interpret radiographic films or ultrasonic scans, relying on years of experience to distinguish defects from noise. These processes are costly, difficult to scale, and vulnerable to human error.

The rapid advancement of Artificial Intelligence (AI), machine learning, and deep learning has introduced transformative opportunities in automated inspection. Convolutional Neural Networks (CNNs), transformer-based architectures, explainable AI frameworks, and synthetic data generation techniques now enable machines to detect, classify, and map defects with accuracy approaching (or in some cases exceeding) that of human experts. More importantly, these systems can operate in real time, process large volumes of data, and provide objective, repeatable results.

This book presents a unified and application-driven exploration of intelligent AI-based defect detection across four major domains:

1. Road infrastructure inspection, focusing on crack detection and real-time mapping.
2. Digital radiographic weld evaluation, emphasizing deep learning-based defect recognition.
3. Arc stud welding defect classification, integrating signal analysis and explainable AI.
4. Ultrasonic testing (UT) inspection, leveraging modern object detection architectures and generative models.

Unlike works that treat these domains separately, this book highlights the shared computational principles underlying defect detection across visual and signal-based modalities. Whether the input is a pavement image, a radiographic X-ray, an arc welding waveform, or an ultrasonic B-scan, the fundamental challenge remains the same: extracting meaningful features from noisy data and making reliable decisions under uncertainty.

DOI: 10.1201/9781003766575-1

A central theme throughout this book is the transition from detection to intelligent decision support. Traditional algorithms often answer a binary question: "defect" or "no defect." Modern AI systems, however, can localize defects, quantify severity, classify defect types, estimate confidence levels, and even provide explainability regarding their predictions. Such capabilities are essential for regulatory compliance, safety certification, and industrial adoption.

Another key contribution of this work is the integration of:

- Hyperparameter optimization techniques, including Bayesian Optimization for efficient model tuning
- Real-time implementation strategies, balancing accuracy with computational constraints
- Explainable AI methods, ensuring transparency and trust in safety-critical applications
- Synthetic data generation (GANs, diffusion models, and transformers) to address data scarcity in industrial environments

Data scarcity, class imbalance, and high annotation cost are persistent challenges in industrial AI applications. By incorporating generative modeling and knowledge distillation techniques, this book addresses practical deployment barriers beyond academic model development.

From a systems perspective, intelligent defect detection requires more than a neural network. It demands a complete pipeline including data acquisition, preprocessing, augmentation, training, validation, performance evaluation, hardware considerations, and deployment strategy. This book presents these components as integrated engineering systems rather than isolated algorithms.

The intended audience includes:

- Graduate students in electrical, civil, mechanical, and computer engineering
- Researchers in AI-driven inspection and non-destructive evaluation
- Industry professionals in infrastructure management, oil & gas, automotive manufacturing, and aerospace
- Practitioners seeking practical implementation guidance for real-world AI systems

By combining theoretical foundations with experimental validation and real-time demonstrations, this book bridges the gap between research innovation and industrial application.

As global infrastructure ages and industrial systems become more complex, the need for intelligent, scalable, and reliable inspection technologies becomes increasingly urgent. AI-driven defect detection is no longer a research novelty, it is a practical necessity. The methodologies presented in this book contribute toward a future in which inspection systems are autonomous, adaptive, explainable, and deeply integrated into the lifecycle management of critical assets.

In the chapters that follow, we move from foundational concepts to domain-specific implementations, demonstrating how artificial intelligence reshapes defect detection across roads, welding, and industrial inspection systems.

Section I

Roads

2 Optimized CNN for Real-Time Road Crack Mapping

M-Mahdi Naddaf-Sh, SeyedSaeid Hosseini, Jing Zhang, Nicholas A. Brake, and Hassan Zargarzadeh

CHAPTER SUMMARY

Pavement surveying and distress mapping is completed by roadway authorities to quantify the topical and structural damage levels for strategic preventative or rehabilitative action. The failure to time the preventative or rehabilitative action and control distress propagation can lead to severe structural and financial loss of the asset, requiring complete reconstruction. Continuous and computer-aided surveying measures not only can eliminate human error when analyzing, identifying, defining, and mapping pavement surface distresses but also can provide a database of road damage patterns and their locations. The database can be used for timely road repairs to gain the maximum durability of the asphalt and the minimum cost of maintenance. This chapter introduces an autonomous surveying scheme to collect, analyze, and map the image-based distress data in real-time. A descriptive approach is considered for identifying cracks from collected images using a convolutional neural network (CNN) that classifies several types of cracks. Typically, CNN-based schemes require a relatively large processing power to detect desired objects in images in real-time. However, the portability objective of this work requires to utilize low-weight processing units. To that end, the CNN's training was optimized by the Bayesian optimization algorithm (BOA) to achieve the maximum accuracy and minimum processing time with minimum neural network layers. First, a database consisting of a diverse population of crack distress types such as longitudinal, transverse, and alligator cracks, photographed at multiple angles, was prepared. Then, the database was used to train a CNN whose hyperparameters were optimized using BOA. Finally, a heuristic algorithm is introduced to process the CNN's output and produce the crack map. The performance of the classifier and mapping algorithm is examined against still images and videos captured by a drone from cracked pavement. In both instances, the proposed CNN was able to classify the cracks with 97% accuracy. The mapping algorithm is able to map a diverse population of surface cracks patterns in real-time at the speed of 11.1 km per hour.

DOI: 10.1201/9781003766575-2

2.1 INTRODUCTION

Societal economic vitality and growth is intimately tied to the state of infrastructure and its ability to safely and efficiently handle the transfer of goods from one point to another. Pavement management, preservation, and rehabilitation strategies are critical components in maintaining the viability of infrastructure and economy over the long term. Roadway networks contain millions of miles of pavements and maintenance operations of these systems cost upwards of $25 billion per year [1]. As part of the maintenance operations, pavement surveys, which include both surface and subsurface assessments, are required frequently to assess the state of the pavement and help to prioritize rehabilitative and preservative action. Moreover, according to The National Highway Traffic Safety Administration, 16% of traffic crashes are produced due to roadway environmental factors mainly by poor pavement conditions [2]. Poor road conditions also lead to excessive wear on vehicles and tend to increase the number of delays and crashes which can lead to additional financial losses [3]. Currently, manual inspection is the most common technique for identifying pavement distress road surveys [4]. Manual inspection can be, however, time-consuming, costly and labor-intensive. Furthermore, during manual inspection operations, human visual error is possible, the operation itself can be unsafe due to the passing of nearby motor vehicles, and the operations may impede traffic flow [5]. To overcome the limitations of manual inspection, automated and/or semi-automated crack detection techniques can be developed to measure, monitor, and map the evolution of the pavement surface and sub-surface structure and distress profile [6]. Semi-automated modern pavement distress mapping or diagnosis techniques need to be non-destructive, cost-effective, accurate, enable data acquisition at high-speed, and relatively user and environmentally friendly [7]. As part of an effort to lower costs and accelerate maintenance operations, transportation departments are prioritizing the development of automated systems profiling systems for pavement distress assessment [8]. There remains a need, however, to develop automated and real-time distress mapping and assessment tools that can provide the end-user with large quantities of information related to the distress-type, geometry, and distress source without manual surveillance-either in-situ or by proxy. The prominent solution of replacing expert inspectors with robots that can automatically gather data and analyze them has been studied or suggested extensively in recent years.

In the literature, automated crack detection works emphasis was set on both using image processing for data analysis and developing automatic method for fast data collection like using robots or vehicles. Advantages like portability, being non-destructive and lane closure avoidance are some of the important aspects of using vehicles for data collection in pavement distress studies, as suggested in many publications [9], [10].

Despite all benefits of automated data collection methods, it leads to vast amount of raw pavement data. Interpreting the raw data needs human expert for analysis and decision making. Regarding the importance of on time maintenance of pavements, it is impossible to process all raw data relying on expert human performance which has led the researchers to develop automatic intelligent algorithms for processing gathered raw data. The utilization of computer vision methods for pavement engineering applications has grown exponentially over the last few decades [11], while many challenges should be addressed to achieve a full and seamless realization due to unwanted and highly variable image noises from random variation of brightness color, camera, and the environment [12]. In recent years many transportation and highway agencies in the US, have become interested in image processing based methods for analyzing collected raw data from highways and roads [13].

Classical image processing algorithms were suggested for pavement distress analysis. Algorithms like edge detection [14], wavelet transforms [15, 16, 17, 18], intensity thresholding, texture analysis, etc. have been well studied [18, 19]. Although numerous suggested classical methods have helped in pavement distress analysis, some drawbacks like being prone to environment noise, not being applicable under all road conditions, being dependent to certain image quality etc. have reduced their robustness against processing varying data. In the recent literature, machine learning based methods, especially deep learning, shows promising results in pavement distress analysis [20, 21, 22, 23, 24]. Unlike classical methods, machine learning based algorithms proved to be more robust for processing different pavement distress images under noisy conditions. Several machine learning methods like neural classifiers [25] and support vector machines (SVM) [26] had been suggested for pavement distress analysis. In a comprehensive review of computer vision based defect detection and condition assessment of asphalt pavement, Koch et al. [11] identified SVM as the most robust machine learning technique for image based pavement distress detection in 2015. Moreover, recently deep learning has become a popular alternative in pavement distress analysis due to its convincing performance over SVM and other methods [27].

Cha et al. [10] created a database with 40,000 images of size pixels and annotated them into crack or intact bins utilizing MatConvNet [28] for crack detection that could achieve 98% accuracy. Gopalakrishnan et al. [29] studied transfer learning on a single layer pretrained neural network classifier for pavement distress detection. They labeled their data as crack or ('1') and no-crack or ('0'). Also, they could gain 90% accuracy by using ImageNet's pretrained VGG-16 as the feature extractor. Dorafshan et al. [30] prepared a database of 1,574 crack and 16,426 without crack images. They compared deep learning and edge detection methods and suggested a combination of both for improving the results in crack detection of concrete. Also, they used AlexNet architecture for feature extraction. Smartphone-based data collection is proposed in Maeda et al. [31], and they have tested several object detection systems like Faster R-CNN, YOLO, SSD, and R-FCN. Gopalakrishnan in [27], has extensively reviewed the most recent deep learning based methods for pavement distress detection. Besides to developing crack distress detection algorithm whether it is based on the classical or deep learning method, some papers have suggested specific platforms for data collection. Due to the vastness of the roadway system, automatic pavement screening is needed. Crasanna et al. [9] has suggested an automated crack detection algorithm, called spatially tuned robust multifeatured, for monitoring concrete bridges and they implement the algorithm on the a robot platform. In [32],

Some had studied an automatic image based road crack detection method and a vehicle based data collection platform is used to collect data from different locations for further processing. Between several data collection methods, vehicle mounted cameras are the most popular one.

The integration of computer vision methods using deep convolutional neural networks (CNN), show exceptional promise for use in crack detection applications, but require many images for the training process [10], [27], [33]. Although utilizing CNNs improves the crack detection accuracy, there are known drawbacks in the conventional approaches for studying the cracks [10]. In the available works [10], [27] whether asphalt pavement or other surfaces, the objective is to distinguish cracked areas of pavement from un-cracked ones that yields to a binary decision with two outcomes: cracked or non-cracked. Due to the proven outstanding performance of deep learning in contrast with other machine learning methods like SVM, Adaboost, and random forest, still some shortages exists. For instance, in [11] although it is proved that CNN has the capability of detecting cracks with high accuracy, authors suggest an optimization for the CNN for the future improvements. Also, using transfer learning for pavement distress detection Gopalakrishnan and et al. [29] suggested to add a feature for evaluating severity of detected cracks which shows possible further improvements for pavement distress detection. In deep learning-based methods, for pavement distress detection, the current focus is on improving the accuracy of neural network for identifying cracks. Also, most of the recent work in this application uses AlexNet, VGG-16 CNN architecture and some transfer learning methods for pavement distress detection task. It is important to consider that most of the mentioned architectures are designed and tested on datasets that do not include pavement distress data. Although in many cases transfer learning is applicable for reducing training time and improving accuracy, the objects in datasets like MNIST, ImageNet, CIFAR-100 etc. do not share similar patterns in pavement applications so using transfer learning with similar CNN architectures is limited.

2.2 METHODOLOGY

As reviewed above, deep learning-based methods for pavement distress detection improves false detection accuracy within non-cracked pavements; whether using different CNN architectures or using transfer learning or pre-trained models. This chapter proposes an approach to geometrically map a surface crack on asphalt pavement using a technique that involves image partitioning and crack geometrical and spatial classification. This technique allows the user to both detect the presence and map a crack on the road surface in real time using raw input images. The work extends the functionality of CNN-based classification techniques, which up to date, are limited to only crack presence detection and do not provide simultaneous geometrical mapping of the object [34], [35]. Crack images are aggregated in the database and indexed according to their orientation and spatial position within a squared partitioned area of the larger raw image file; which then allows the position and orientation to be estimated heuristically using thirteen unique categories. By applying this approach, instead of predicting crack position in each frame by a marginal error that depends on the searching window, we not only are able to detect and classify the cracks, but also map the crack and avoid errors caused by the searching window.

Instead of relying on pre-designed CNN architectures, we have proposed an optimized architecture for the CNN and hyperparameters within the pavement distress detection task. Also, rather than taking the approach in the crack detection task that is focused on detecting crack from non-crack; which introduces some false positive after classification, in this work we propose a heuristic algorithm CMA which is able to regenerate crack shape automatically. Alternatively, in this work we propose a method for mapping a crack's shape or analyzing the results based on the pattern of a crack in an entire image or video frame that could yield a smoother crack map by eliminating smaller cracks or errors. Moreover, the proposed method is a general concept for automating the road surface crack analysis that can be later adapted with several highway agencies protocols like American Association of State Highway and Transportation Officials AASHTO (PP67-10 and PP44-00) or Mechanical-Empirical Pavement Design Guide MEPDG [36]. In other words, by changing the camera parameters like sensor size, focal length, lens type, distance of camera from pavement etc. it is possible to achieve the minimum deficiency length that is considered crack in several protocols.

In the proposed scheme, each image is first partitioned into 300 equal square tiles. Then, a CNN is developed and trained that classifies the cracks in the tiles to predefined categories. Since the categorization of the cracks is conducted tile-by-tile, the resulted map may show discontinuities at the borders of the tiles. To mitigate such discontinuity errors, a heuristic real-time crack mapping algorithm (CMA) is introduced. The CMA processes the classification results and, based on the regional crack's information, it modifies the current segment that yields to a unified and continuous map of cracks on the road surface image. Further, the CMA has the ability to eliminate small cracks or false positive objects isolated in one partitioned tile that is continuous over multiple partitioned tile. Since the objective of this chapter is to map the crack real-in-time in image frames from a streaming video, the CNN hyperparameters (HPs) had to be optimized so the processing time and the classification error for the input images are minimized simultaneously. To that end, a Bayesian optimization algorithm (BOA) was utilized in lieu of trial and error methods. Experimental results show that the CNN installed on a portable computer can process 5 frames per seconds providing the ability to map one band of a road real-in-time at the speed of 11.1 km per hour.

The primary objective of this work is to use real-time images to map cracks on the surface of an asphalt pavement. A crack is defined as a mechanical or thermal strain-induced separation of material. This material separation allows moisture to infiltrate the pavement structure internally; leading to premature failure or accelerated deterioration. Cracks are classified by their geometric orientation, source, width, and concentration per unit length or area. Only cracks visible and distinguishable to the naked eye are considered in the distress survey.

As illustrated in Figure 2.1, the proposed crack mapping scheme is comprised of three stages: database preparation, training and optimizing, and real-time crack

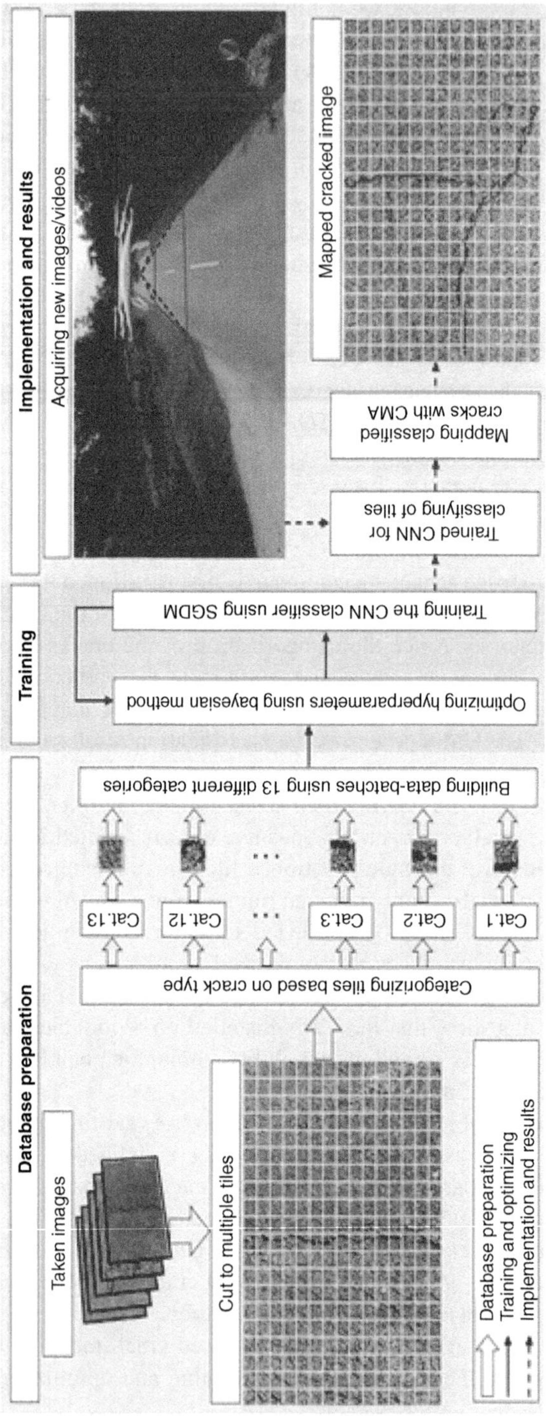

Figure 2.1 Flowchart for Crack detection and mapping.

mapping. To prepare the database, a descriptive approach was taken to categorize a given crack based on its relative position and geometric orientation within the image. This approach labels the cracks based on their geometric orientation. Multiple images of cracked asphalt pavements were gathered from the field and each image was sub-divided into smaller tiles (Tls). The Tls were classified into 12 different categories based on whether the Tl contained a crack. If a crack is detected, it is further classified by crack position and orientation, i.e., horizontal, vertical, diagonal. Tls that did not contain a visible crack, but rather contained objects like grass, shadows, patched cracks, pavement markings, general uncracked pavement, etc., were binned into a 13th category. A CNN is then used to learn the unique features contained within each Tl, and then each Tl was classified into a specific category (1-13) based on crack presence, position, and orientation. To further refine the process, BOA is used to objectively and systematically achieve optimal HPs by selecting optimal initial learning rates and momentum; which served to significantly reduced training time operations that have been shown to be tedious when tuned manually [12], [10]. The trained CNN is used for real-time crack mapping using video frames of a cracked roadway surface that were received from a camera installed on an aerial vehicle. The classified Tls are then sent to the CMA which maps the cracks in real-time. The CMA is designed to enhance the decisions made by the CNN by eliminating cracks found only within one isolated Tl and requiring the crack maps to be contiguous across multiple Tls to increase the smoothness and accuracy of the mapped crack field.

2.3 EXPERIMENTAL STUDY

2.3.1 DATABASE PREPARATION

To develop and evaluate the CMA, 1500 images of cracked asphalt pavement surface were collected to prepare a database to train the CNN. Images were taken by a FLIR E5 camera with a $55^\circ \times 43^\circ$ field of view and 640×480 pixels resolution. The camera was placed between 1.5m to 2.5m above the road with a vertical line of sight and $\pm 10^\circ$ of tilt error. Each image was then divided into 300 equal Tls containing 32×32 pixels. Each Tl was then virtually divided into 9 equal blocks, as depicted in Figure 2.2(a); where the hashed blocks indicate the range of possible crack locations within a given bounded region; which is uniquely defined for each category. Figure 2.2(b) shows how an actual pavement surface image is aligned with a given category. Category groups $\{1,2,3\}$, $\{4,5,6\}$, and $\{7,\ldots,12\}$ represent horizontal, vertical, and diagonal cracks, respectively. The non-cracked category (13) is shown in Figure 2.2(c).

Based on the established definition of categories, a total of 6,695 Tls were handpicked from the 1500 images that fit into one of the 13 categories. For representing data, CIFAR-10 [37] layout was used that yielded to 13 different data-batches equal to the number of crack categories. Each data-batch contains similar manually annotated Tls that represents one of the 13 categories along with an assigned label $Lb \in \{1,2,...,13\}$ indicating the Tl's category number. Then, 90% of each data-batches were randomly taken along with their labels to form the training set and the

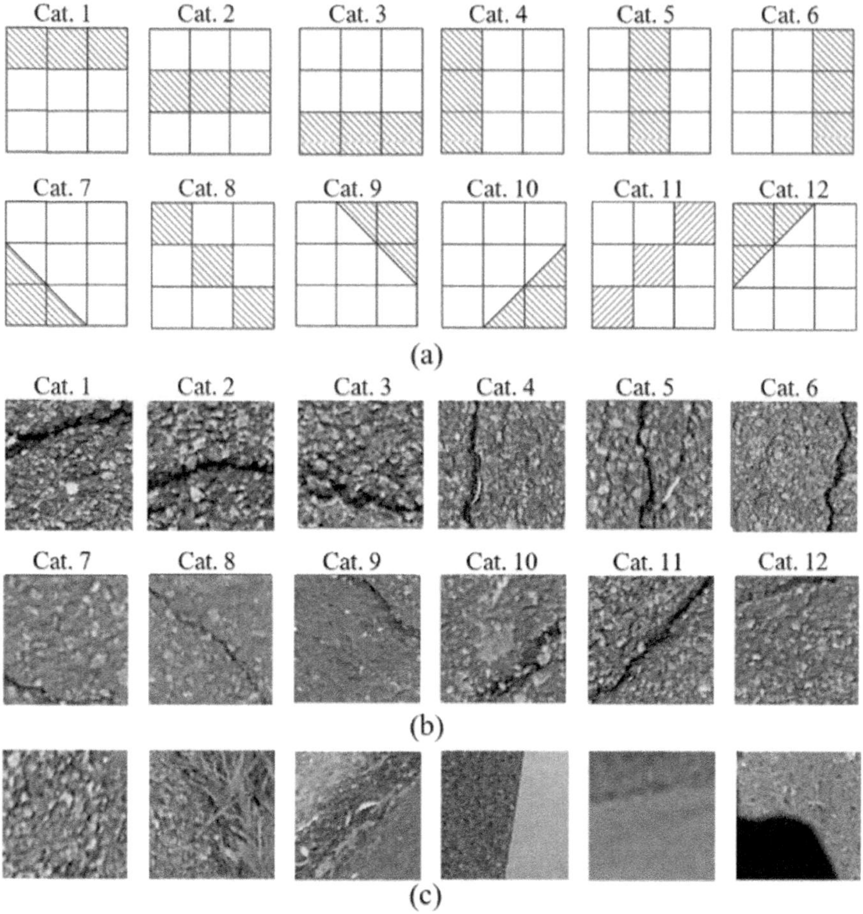

Figure 2.2 (a) Categorization method based on the position of a crack in each tile, where indicates a defective section, (b) Sample of tiles in 12 categories, (c) Category 13th sample images of non-cracked tiles.

other 10% was used for the test set. A test set is then used to measure the accuracy of the trained CNN.

2.3.2 CNN ARCHITECTURE

Figure 2.3 illustrates the overall CNN architecture-input, multiple convolutional layer, followed by a batch normalization layer, ReLU layer, max-pooling layer, a fully connected (FC), SoftMax layer and output layer for the classification task.

The first layer is the input layer that receives the input image Tls for classification. The crack features in each data batch were extracted using multiple convolution layers. This layer consists of various sets of neurons whose weights and biases will

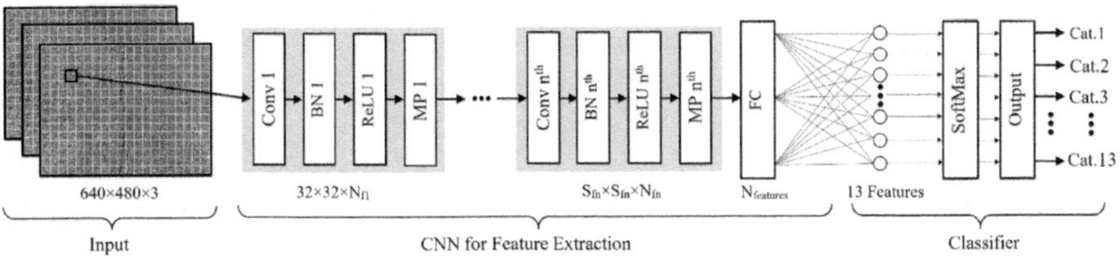

Figure 2.3 Network architecture that is used for crack detection. The number below each layer is indicated the size of each layer. Conv: Convolution layer, BN: Batch Normalization layer, MP: Max-pooling layer, and FC: Fully Connected Layer.

be updated relative to the crack features. In the convolutional layer, the neuron input consists of small sectors from the previous layer that is called the filter (kernel). The size of the filter, S_f, can be tuned from $1x1$ pixels up to the size of the input image. In the convolution layer, the filter moves along the input and builds a convoluted feature map. To increase the number of feature maps, multiple filters should be used, and each filter has different weights and biases to be able to extract various features of the image. The stride (amount of horizontal and vertical movement of the filter on the input per convolution) is set to 2 pixels.

After the convolution layer, a batch normalization layer is used for reducing the CNN sensitivity to initial HPs values and decreasing training processing time. Following the batch normalization layer, a Rectified Linear Unit (ReLU) activation layer is added to apply a zero threshold to all negative values in the batch normalization layer, that means, the inputs from the previous layer b go through $\max(0,b)$. The max-pooling layer downsamples the input by dividing it into rectangular pooling regions to compute the maximum of each region of gathered feature matrices. After designing the feature extractor, the Fully Connected (FC) layer is used to map the features matrix in the last layer in the form of a $1xs$ vector, where $s = 13$ is chosen equal to the number of categories in the database. For representing the probability distribution over multiple classes in the output of a classifier, a generalized model of binary logistic regression classifier (SoftMax function), is utilized after the FC [38, 39]. Considering the input of the SoftMax function as a sample tile that belongs to one of 13 categories, $Tl \in Cat_j$ where $j \in \{1,...,13\}$, then the category prior probability [40] is defined as $P(Cat_j)$ which shows the probability of $Tl \in Cat_j$ and conditional probability as $P(Tl,\overline{\phi}|Cat_j)$, where $\overline{\phi} = [\overline{w},\overline{b}]$ is the parameter vector that consists of weights ($\overline{w}$) and biases($\overline{b}$). The SoftMax function is described as:

$$S_j(Tl,\bar{\phi}) = P(\mathrm{Cat}_j|Tl,\bar{\phi}) = \frac{P(Tl,\bar{\phi}|\mathrm{Cat}_j)P(\mathrm{Cat}_j)}{\sum_{m=1}^{j} P(T_l,\bar{\phi}|\mathrm{Cat}_m)P(\mathrm{Cat}_m)} = \frac{\exp(r_j(T_l,\bar{\phi}))}{\sum_{m=1}^{j}\exp(r_m(Tl,\bar{\phi}))} \tag{2.1}$$

where $r_j(Tl,\overline{\phi}) = \ln(P(Tl,\overline{\phi}|Cat_j)P(Cat_j))$ and S_j is a probability distribution as the SoftMax function output; where $0 \leq S_j \leq 1$ and $\sum_{j=1}^{13} S_j(Tl,\overline{\phi}) = 1$.

Following the SoftMax function, the classification output layer (cross entropy function) is used to assign each input to one of the $n = 13$ mutually exclusive categories using the loss function shown in the following:

$$l(\bar{\phi}) = -\sum_{i=1}^{p}\sum_{j=1}^{n} d_{ij} \ln S_j(Tl_i, \bar{\phi}) \tag{2.2}$$

where p is the number of samples and $d_i j$ is a matrix that shows with what probability the ith sample of Tl belongs to j^{th} category.

2.3.3 TRAINING

Stochastic Gradient Descent with Momentum (SGDM) was used to train the CNN for classification. This method updates CNN's weights and biases to minimize the loss function that measures the difference between true classified and false classified Tls. The SGDM uses a subset of training data (mini-batch). The gradient derived from the data within the mini-batch is used for updating the weights and biases. Each update to the weights and biases is defined as one iteration. The gradient descent update law is described as:

$$\bar{\phi}_{k+1} = \bar{\phi}_k - \lambda l(\bar{\phi}_k) + \eta(\bar{\phi}_k - \bar{\phi}_{k-1}) \tag{2.3}$$

where subscript k represents the iteration number, the initial learning rate is $0 < \lambda < 1$, $\overline{\phi}$ is a vector that contains the weights and biases, $l(\overline{\phi})$ is the loss function and $0 \leq \eta \leq 1$ is the momentum, which defines the level of contribution from the previous step. For λ values close to 0, the learning process is slowed, and values close to 1 lead to either diverging or suboptimal weights. Moreover, to prevent overfitting of CNN during the training process, L2 regularization [39, 41] is utilized as follows:

$$l(\overline{\phi}_{k+1}) = l(\overline{\phi}_k) + \frac{\tau(\overline{w}_k^T \overline{w}_k)}{2}, \tag{2.4}$$

where τ is the regularization factor. To both prevent over-fitting and feature memorization, and improve the generalization of the SoftMax classifier during the training process, a modified data augmentation procedure is used during each iteration [38]; where the Tls were translated randomly in the horizontal and vertical directions by a maximum of ± 4 pixels. It is noteworthy that the Tls cannot get flipped or rotated since the classification process is dependent on crack orientation.

2.3.4 NETWORK HYPERPARAMETERS (HP) OPTIMIZATION

HPs in the proposed CNN architecture and SGDM are the filter size S_f, number of filters N_f, the number of CNN layers N_D, η, τ, and λ. The search range for HP was defined as $N_D \in \{1,2,\ldots,15\}$, $S_f \in \{1,2,\ldots,12\}$, $N_f \in \{1,2,\ldots,100\}$, $0 \leq \eta \leq 1$, $0 \leq \tau \leq 1$, and $0 < \lambda < 1$. The possible values for N_D, S_f and N_f are integers, and for η, τ, and λ are logarithmically spaced values between 0 and 1. The classification error is the number of miss-classified Tls by the classifier (SoftMax). The objective of optimization is to find optimal values for the HPs such that the classification error is minimized. So, the objective function can be considered a function with HPs as the input and the classification error as the output. Modeling of this objective function

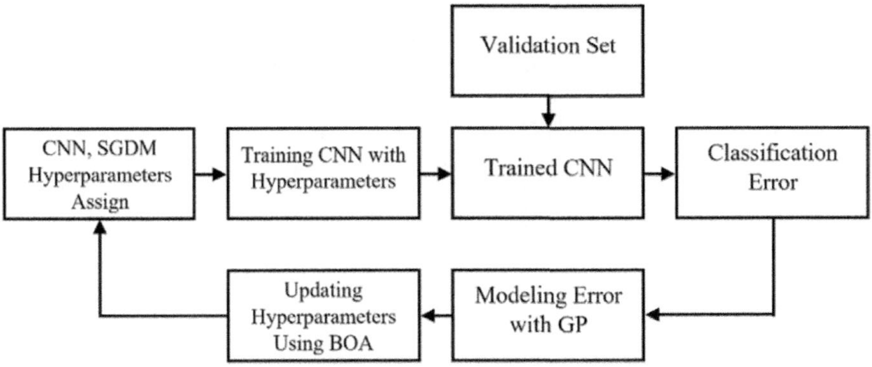

Figure 2.4 Block Diagram of optimizing HPs using BOA.

is algebraically complicated and computationally intensive. The BOA is capable of performing optimizing the HPs to minimize the classification error, while the objective function is considered as a black-box [42]. To perform the BOA, a validation set was defined that consists of 10% randomly selected Tls from the training set. The inputs of the objective function are training set and validation set.

As shown in Figure 2.4, the objective function trains the CNN and returns the classification error on the validation set. By modeling the calculated error using Gaussian process (GP) as mentioned in [43] and in multiple iterations z, where $z = \{1, 2, \ldots, 100\}$, the BOA finds the optimal values for HPs that minimize the classification error. The kernel function that was used for GP is the Automatic Relevance Determination (ARD) Matérn 5/2 in [44]. In addition, the acquisition function ($q_z(\text{HP})$) that is used for the GP is the Expected Improvement function $\mathbb{E}(\cdot)$ [45], as follows:

$$q_z(\text{HP}) = \arg\max_{\text{HP}} \mathbb{E}\left(\max\left\{0, f_{z+1}(\text{HP}) - f_z^{\max}(\text{HP})\right\}\right) \tag{2.5}$$

where $f_z^{\max}(\text{HP})$ is the current maximum observed value for the objective function. The next estimation for maximizing the objective function is obtained by using the acquisition function. The GP posterior is updated in each iteration using Equation 2.5:

$$P(f_z|u) = \frac{P(u|f_z)P(f_z)}{P(u)}, \tag{2.6}$$

where $u = \{(\text{HP}_z, f_z), z = 1 : 100\}$.

The extrema of $f_z(\text{HP})$ was obtained numerically at sampled values of the function. A closed form expression of the objective function is not required within the BOA mathematical structure [46]. The objective function and acquisition function for two of SGDM HPs, i.e., η and λ during the optimization process are shown in Figure 2.5. As depicted in Figure 2.5(a), the observed points are demarcated by blue dots ($f_z(\text{HP})$), the model mean that is obtained from the observations is depicted as the red surface, and subsequent evaluation point addition is demarcated with a black

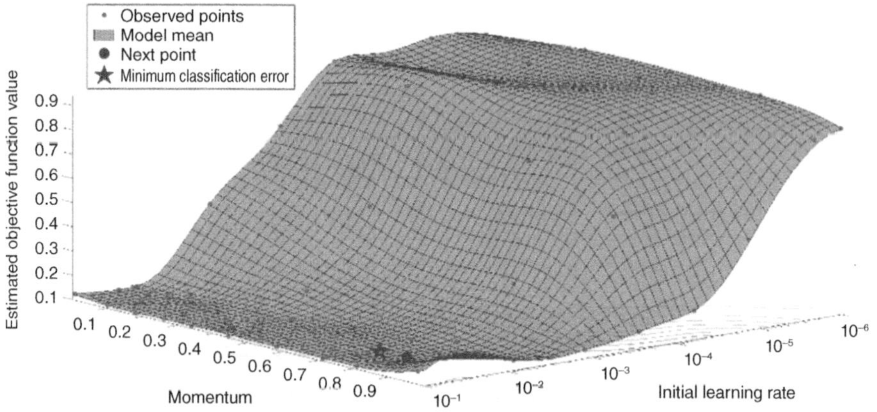

(a)

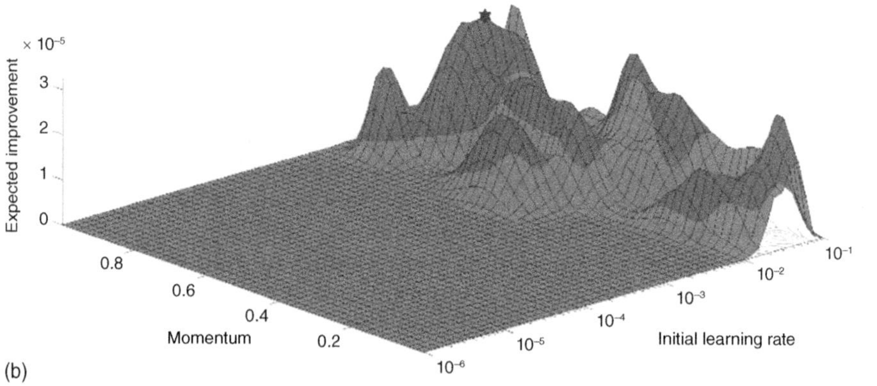

(b)

Figure 2.5 (a) The observation function model.(b) Acquisition function for two parameters of the SGDM. The starred point is the calculated optimal values at the 69th iteration for λ and η.

dot. Moreover, Figure 2.5(b) illustrates the acquisition function. The objective function is shown to reach a minimum at the 69th iteration; this point is demarcated with a black star. Figure 2.5(b) shows the maximum feasible value that is generated upon minimizing the classification error.

The total number of iterations was set to 100. Each iteration calculates the classification error among 600 randomly selected Tls from 13 data-batches. Figure 2.6 represents the estimated (expected) improvement in each iteration and the calculated improvement during optimization. The BOA was evaluated statistically using the Wald method [47] by representing the images in the test set as independent events with a known probability of success. The number of misclassified images were represented with a binomial distribution. By applying the trained CNN with optimized HPs on the test set and computing the number of correctly classified Tls, the test

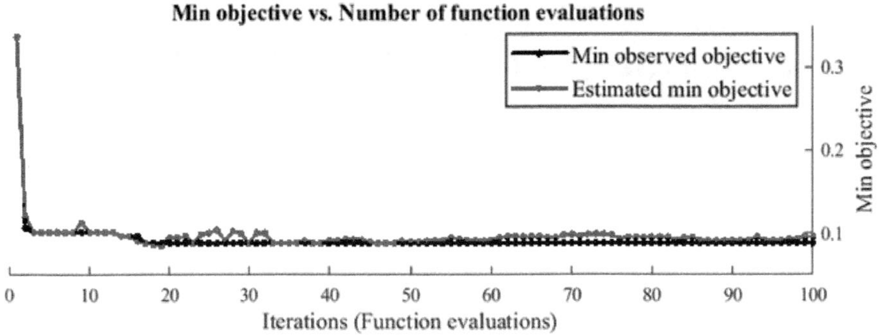

Figure 2.6 Evolution of minimum observed objective and estimated minimum objective in total 100 iterations, where the minimum calculated value is 0.07.

error E_t is defined as follows:

$$E_t = 1 - \frac{1}{b}\sum_{i=1}^{b} Ts_i, \tag{2.7}$$

where Ts and b are the number of correctly classified Tls, and the total number of Tls in the test set, respectively. Note to evaluate the trained CNN performance on the test set without exposing the CNN to the optimization process, E_s is used to obtain the standard error. This approach helps to increase the optimization speed. The standard error is represented as follows:

$$E_s = \sqrt{\frac{E_t}{Ts}(1 - E_t)} \tag{2.8}$$

Moreover, as the target of this research, to obtain a $\pm 3\%$ error margin, a confidence interval of 97% is defined to calculate the generalization error E_G defined as:

$$E_G = E_t \pm 0.97 E_s \tag{2.9}$$

The final HPs value for the CNN were $N_D = 4$, $S_{f1} = 5$, $S_{f2} = 5$, $S_{f3} = 6$, $S_{f4} = 2$, $N_{f1} = 119$, $N_{f2} = 119$, $N_{f3} = 108$, and $N_{f4} = 96$. Also, the optimized values for SGDM were $\lambda = 0.0005145$, $\eta = 0.040069$, and $\tau = 0.002859$.

Applying the optimal HP values to the CNN and SGDM yields a CNN with 24 layers and 96.67% accuracy in 10 epochs as it is shown in Figure 2.7(a). Moreover, the minimized value for the loss function was 0.033 in 10 epochs, as shown in Figure 2.7(b). The generalized error E_G interval for the test set was $[0.0476, 0.0184]$. Figure 2.8 illustrates 24 randomly selected output of the fourth convolution layer, indicating that the features of Tls extracted by the CNN. Extracted features of the position and orientation of a crack in each Tl can be easily recognized in Figure 2.8. Figure 2.9 depicts the confusion matrix for the test set, which is obtained based on the CNN that is trained using final values. As shown in Figure 2.9, the highest confusion is between

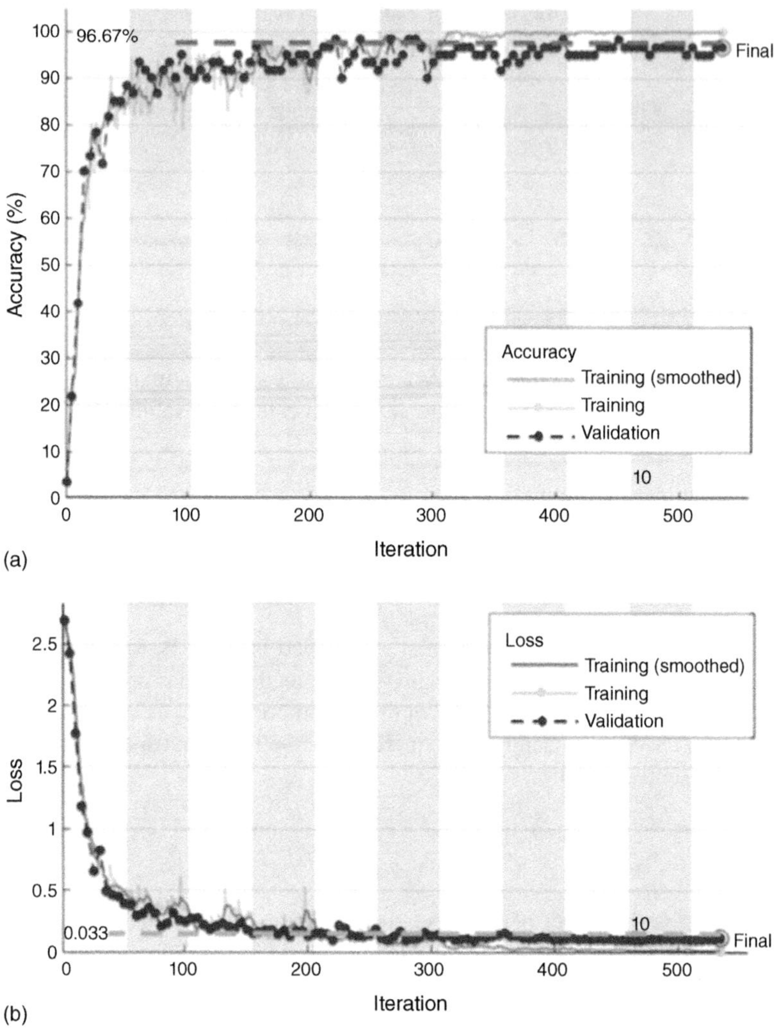

Figure 2.7 (a) Training CNN with optimized calculated value is reached 97% accuracy after 10 epochs. (b) Loss function output during training, the error value is 0.033 after passing 10 epochs.

the two categories 5 and 6. The reason is that, from Figure 2.2(a), categories 5 and 6 both represent vertical cracks in the middle and right side of Tl, respectively, which intrinsically leads to a higher probability of misclassification. 24 randomly selected Tls from the test set that were selected for testing the optimized trained CNN are shown in Figure 2.10 along with the percentage probability that they belong to each

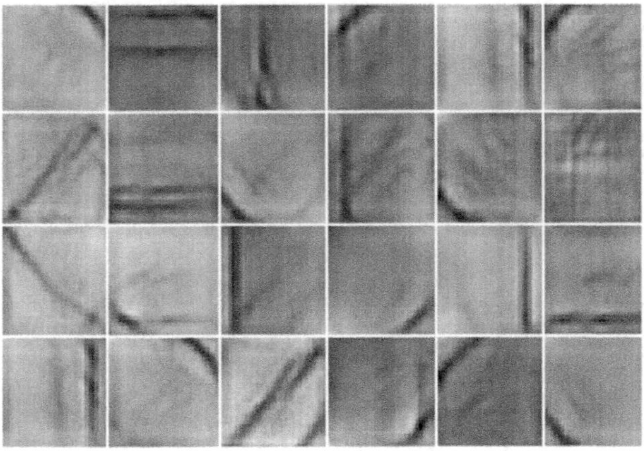

Figure 2.8 24 randomly extracted features in the last convolution layer.

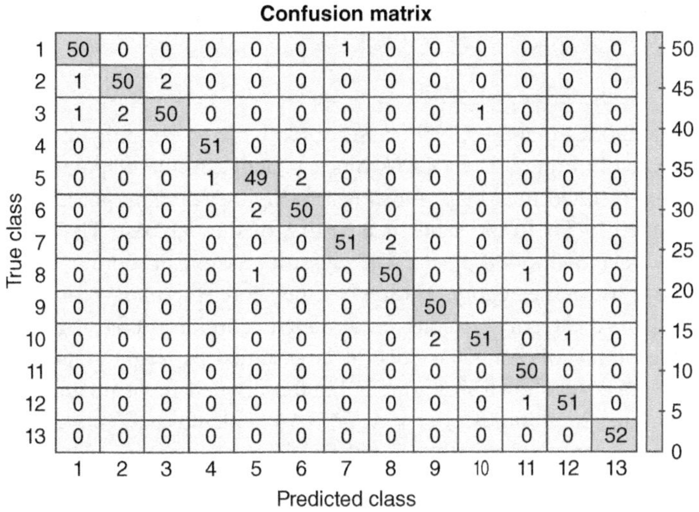

Figure 2.9 Confusion matrix between all 13 categories of the test set.

category. The trained CNN in this section will be used to map the cracks with the CMA described in the following section.

2.4 Real-Time Crack Mapping Algorithm (CMA)

In this section, the proposed real-time Crack Mapping Algorithm is discussed. So far, a CNN is trained that classifies the cracks in a Tl. However, if the CNN is directly used to map a crack, the resulted map will not be continuous. The discontinuity is the result of the classification errors, due to the size of the tile blocks and the limited

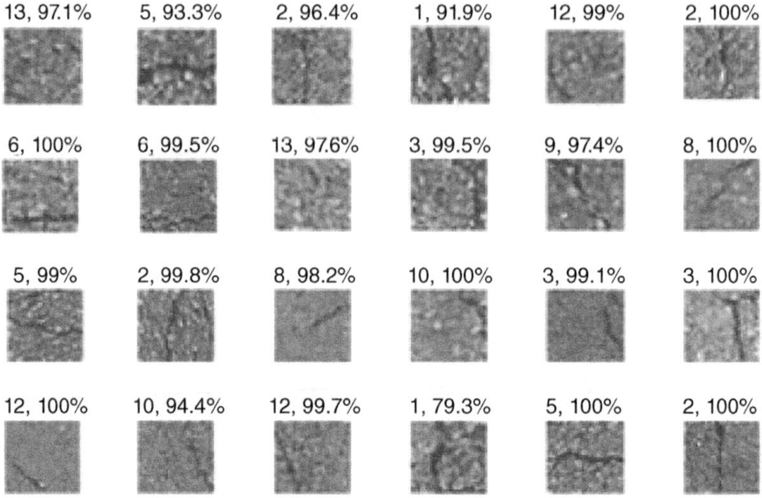

Figure 2.10 24 randomly selected tiles from the test set. Category and its confidence are available on top of each tile.

number of categories that the CNN classifier recognizes. To mitigate this problem, as it is indicated in Figure 2.1, a CMA block is added to the mapping scheme that smoothens the resulted final map. As shown in Figure 2.3, an input image or video frame is divided to Tls in the same size as that of the database Tls assuming that input images have 640×480 pixels. The classifier assigns a score to Tls related to their similarity to each category. The higher the score, the more probable that Tl's crack belongs to a category. Among all assigned scores for a Tl, if a score is less than a defined threshold value of 85%, it is assumed as a non-cracked Tl that is the 13th category. As shown in Figure 2.11, the divided input image has 15 rows $r = A, \ldots, O$, and 20 columns, $c = 1, \ldots, 20$. As depicted in Figure 2.12, for each crack category, a raw mapping plot was defined. The raw mapping plot is a collection of straight-line segments that estimates crack position and orientation based on its classification. This mapping will not be interconnected between tiles which leads to two immediate deficiencies: isolated cracks and non-connected cracks which lead to an overall mapping error.

To improve the crack connectivity between tiles, without refining the pixel dimensions of the Tls and increasing the number of classification categories, a CMA block was created to "tie" the crack line segments between neighboring tiles. This procedure is described as follows: if a Tl has any common side (S_i) or corner with other, Tls it is called neighbor Tl. Each Tl has at least 3 neighbors at the image corner and at most 8 neighbors in the image interior. An isolated Tl is a Tl with a detected crack, but no neighbor Tls have detected cracks. The CMA begins to map the input image in real-time by scanning from row A from left to right. To map each row, the CMA processes right, left and upper neighbors. While scanning a row, isolated Tls

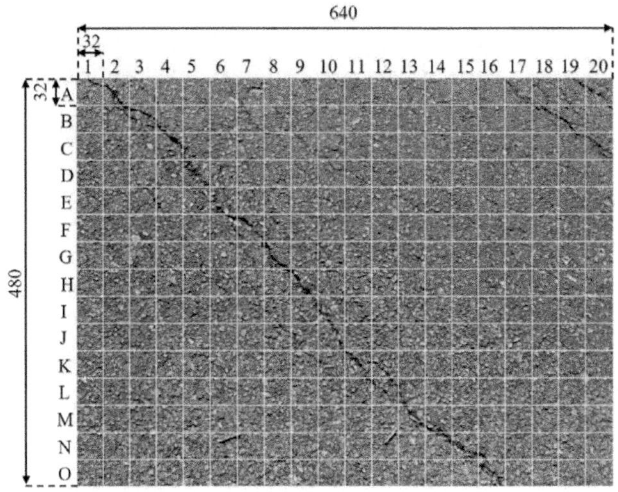

Figure 2.11 Montaging all 300 tiles.

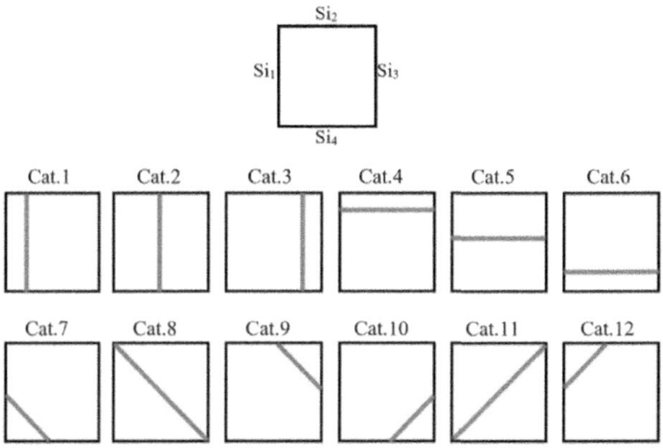

Figure 2.12 Raw plot based on the position of each crack in a single tile for all 12 categories that contain a crack.

are discarded as non-crack Tl. On the other hand, if two or more crack neighbor Tls exist, the raw linear segment plot will be translated vertically or horizontally until the raw mapping segments in the Tl and neighbor Tl intersect. As it is depicted in Figure 2.13(a), the cracks in Tls are mapped by a linear segment and the CMA starts scanning from row A (Left to right). The tile A2 has a neighbor with a shared side on A2 and B2. The CMA translates the start of linear segment vertically downward until intersecting the end of the segment in A2, as shown in Figure 2.13(b). After scanning row A, the next row B is then scanned by the CMA. Tile B2 has a detected crack and detects a crack in the neighbor tile A2. As depicted in Figure 2.13(c), the CMA then translates the end of the line segment within B2 horizontally to the right

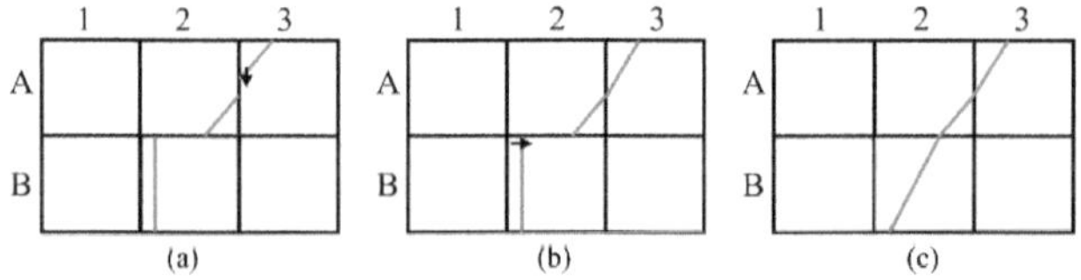

Figure 2.13 Applying the CMA, to make the crack mapping unify.

until intersecting the start of the line segment in A2. This procedure continues until every row in the image has been processed.

Figure 2.14 shows the connected CMA modified mapping plots (red lines) overlaying both crack pavement image and raw mapping plots (blue lines; without segment connection). An isolated crack (blue line) is detected in tile C7 and classified into category 7. No neighbor tiles to C7 have a detected crack. More isolated cracks are detected in C7, E9, H17, and K8. The isolated cracks are eliminated within the CMA. The surviving linear segments are those that have neighbor Tls which are shown from A9, B9, B10, C9, D9… to O13. By defining the neighbor Tl, the mapped plots from A9, B9, B10, C9, D9… to O13 form a pattern which can map the underneath crack. The proposed method is not limited to a certain vertical distance of the camera from the road. By having the camera's parameters like focal length, sensor size etc., the pinhole camera model in [48] can be used to find the real dimensions of the road surface in each image. This information is required to determine the size of the crack in each image. As stated in section 3 images were taken from distance of 1.5 to 2 meters above the cracked asphalt. For instance, for the camera used to collect the images in this work, a 640 × 480 pixels image would cover a 1036 mm × 915 mm to 1280 mm × 1097 mm block on the road. In that case each Tl would cover a 51.8 mm × 61 mm to 64 mm × 73.1 mm area.

Therefore, considering the isolated Tls that are eliminated by the CMA lead to exclusion of cracks with maximum length of 80.02 mm to 97.1 mm. On the other hand, crack's width is one of the other requirements by different protocols for defining severity of deficiency. As an instance in AASHTO protocol PP44-00 and PP67-10 has three different levels for measuring the damage severity. Level 1 is defined as cracks with width of less than 3 mm, level 2 referred to cracks with width between 3 mm and 6 mm and level 3 is cracks with width of greater than 6 mm. Moreover, three major types of cracks in most protocols are longitudinal, transverse and alligator cracks [36], [49]. Although converting length and width of cracks to the percentage of lane area is a straightforward task for transverse and longitudinal cracks, there is no certain way for applying the same method for alligator cracks. Hence some protocols like HPMS, LTPP and PP44-00 are focusing on transverse and longitudinal cracks and considering alligator cracks as a combination of those two types. With the mentioned equipment that is used in this chapter, it is possible to detect cracks with minimum width of 2 mm and maximum width of 30 mm in images that are taken from 1.5 to 2 meters of the pavement surface. The proposed algorithm has no limitation over the length of crack. Moreover, the proposed algorithm is capable

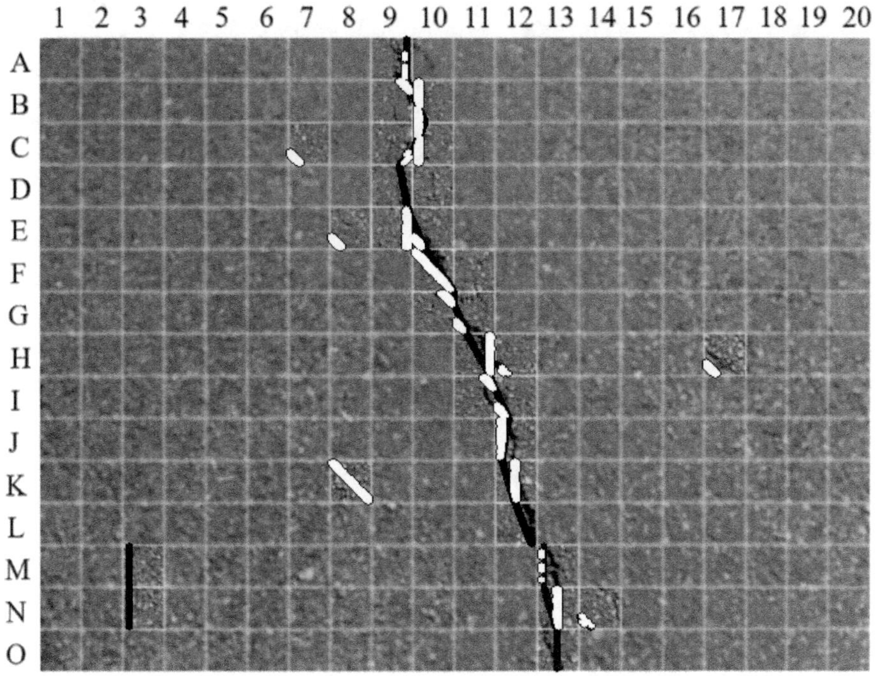

Figure 2.14 Blue lines show the raw mapping of the cracks. The crack map is not unified, and there are isolated mapped cracks available in the image. Red line is the improved mapping the cracks using the CMA, isolated cracks were eliminated, and crack's map is unified and smoothed.

of detecting all three types of cracks that is defined widely accepted protocols.

2.5 EXPERIMENTAL RESULTS

In this section the test results of trained CNN are presented in two subsections: i) the performance of the algorithm was tested on several single images with various crack shapes and ii) a real-time mapping evaluation was done on a captured video that is obtained from a randomly selected pavement roadway. The test images from the training and testing database were not used, were not filtered or modified, and were taken under varying light condition and camera position. Also, in this chapter, MATLAB® was used for training and optimizing the CNN and implementing the CMA algorithm.

A) Single Image Mapping Evaluation

As depicted in Figure 2.15, samples of three images having typical types of cracks with vertical (Figure 2.15(a)), horizontal (Figure 2.15(b)), and diagonal

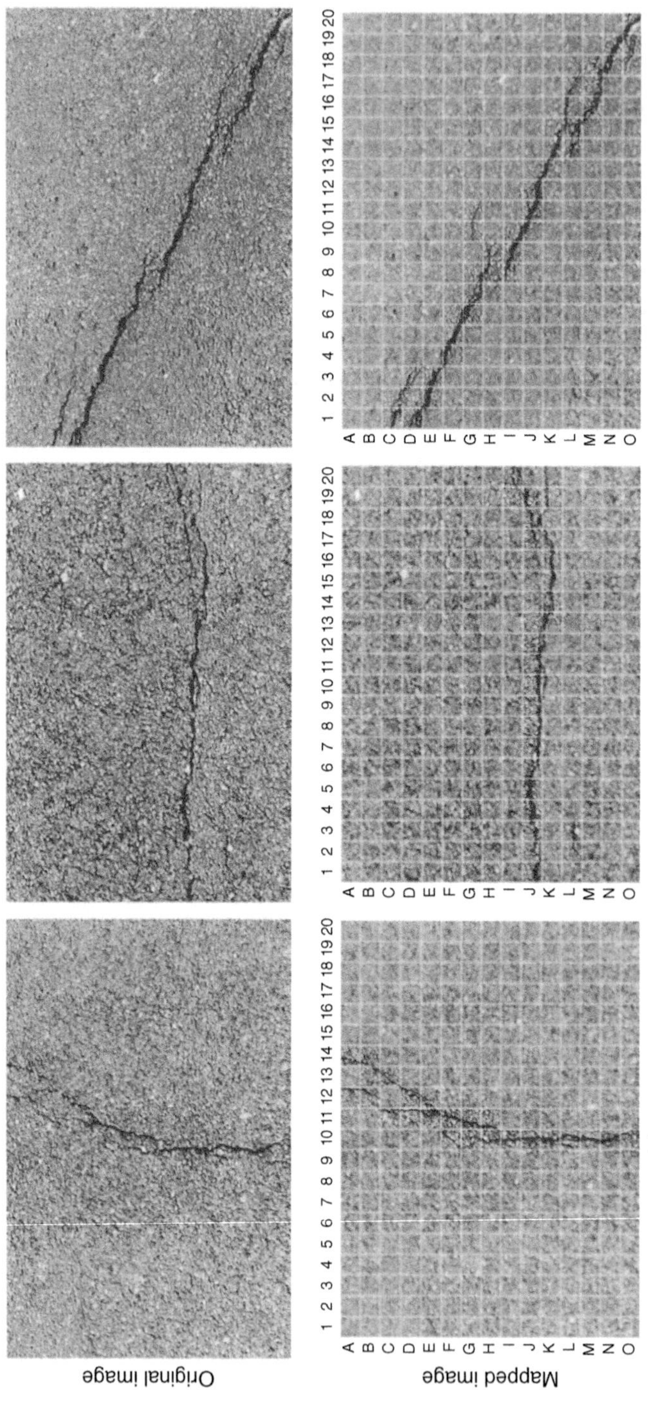

Figure 2.15 Three images with major types of cracks, (a) vertical, (b) horizontal and (c) diagonal cracks are used to test the crack mapping algorithm.

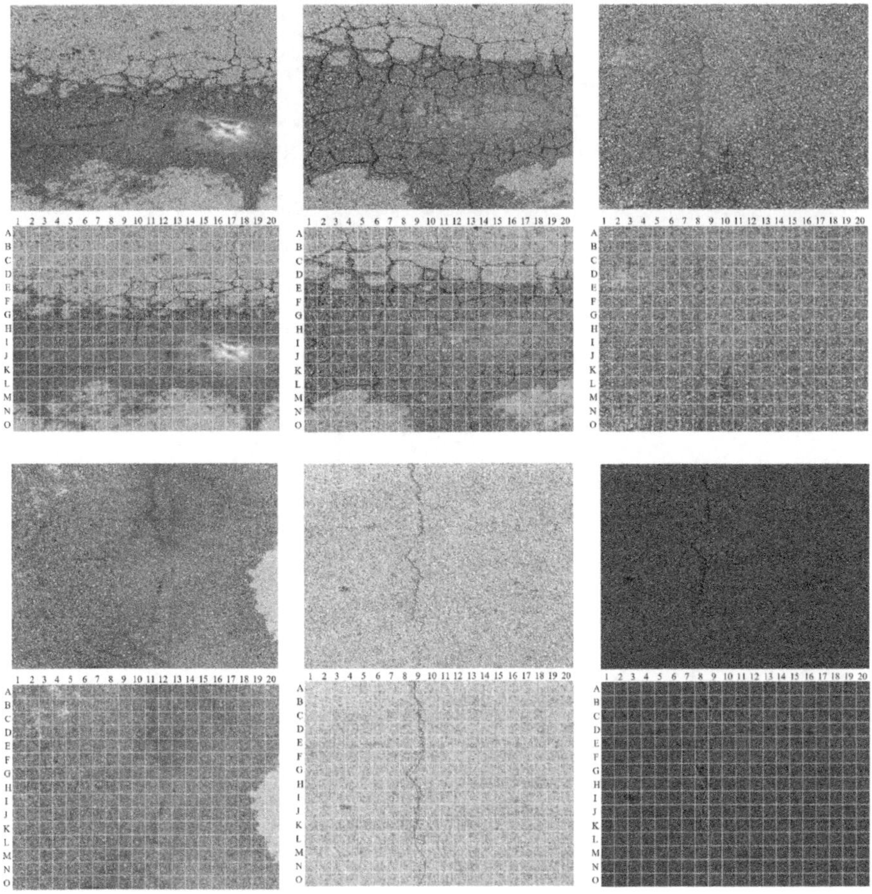

Figure 2.16 Performance of the algorithm on multiple cracks under different conditions (a), (b), (c), and (d) wet, (e) bright and (f) dark.

(Figure 2.15(c)) orientations were used to test the CMA. There are no isolated tile cracks, and the connected mapped cracked covers the main crack with substantial accuracy.

Moreover, to test the algorithm's performance under different weather and illumination conditions multiple cracks were selected and pictures in different situation. Figure 2.16(a), (b), (c), (d), (e), and (f) shows the wet, bright and dark conditions, respectively. The CMA could map cracks regardless of these road condition.

B) Real-Time Mapping Evaluation

Also, to evaluate the CMA performance in a real-time manner, a video was captured using a DJI Phantom 4 drone with mounted camera. The CMA was implemented on

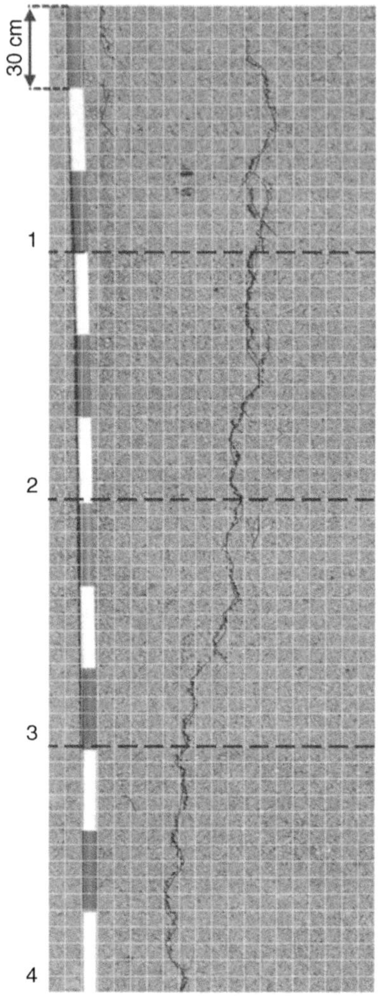

Figure 2.17 4 Processed frames with detected and mapped cracks. Mapping algorithm was able to separate all of 4 significant cracks and map them.

the captured video's image frames. The drone was set to an altitude of 2.5 m from the ground. The camera was perpendicular to the ground, and the video quality was set to 640 × 480 resolution and 120 fps. Since the drone was flown with a slow speed above the crack, and due to the high frame rate, the frame interval for processing was set to 60 frames. A ranging rod with a 30.48 cm with orange and white bands was set beside a road crack, and the drone was flown 3.6 meters in the forward direction. The laptop that was used for processing the captured videos was Alienware 15 R3 with NVIDIA GTX 1070 GPU and Intel Core i7 processor. As it is shown in Figure 2.17, the

Figure 2.18 The real-time video of CMA is available at this link: https://youtu.be/SGh03B5EEpk

algorithm was able to detect and map deep cracks and avoid oil spills on the surface road. Referring to Figure 2.17, each frame was taken every 0.2 milliseconds. A total of 5 frames could be processed per second which covers 0.6 m of the road pavement. This indicates that the maximum speed of the real-time mapping with the current hardware is 11.1 km/h. This speed, however, can be increased with the advent of lighter and more powerful graphical and processing units. Figure 2.18 shows a video obtained from the drone footage with real-time crack mapping segments produced from the CMA.

The proposed algorithm in this work improves upon existing work [10] by integrating crack detection with a crack mapping using image segmentation and classification within a CNN architecture. In addition, the optimized CNN architecture proposed here uses a significantly lower number of filters in the convolution layer (256) leading to reduced computational demand in both CNN training and real-time processing. As mentioned in section 2.6, the BOA is used to compute the HPs. To verify the fact that the selected optimal values maximize the CNN accuracy, during the training process, all HPs were perturbed by a $\pm 5\%$, $\pm 10\%$, $\pm 20\%$ and $\pm 30\%$ white noise. As it is shown in Table 2.1, perturbing the HPs by $\pm 5\%$ decreases the accuracy by about 1%, while as it increases to $\pm 30\%$ the accuracy decreases by at most 8%.

Table 2.1
The effect of changing hyperparameters on model accuracy.

Changing the HPs %	Accuracy %
+ 5%	96.0
- 5%	95.9
+ 10%	95.2
- 10%	95.7
+ 20%	94.6
- 20%	93.1
+ 30%	92.4
- 30%	89.0

2.6 CONCLUSION

In this chapter, an algorithm for mapping road cracks in real time using convolutional neural networks was proposed and tested. Authors gathered the database for this work, and due to limited available resources, the size of the database was limited to 6695 images. The convolutional neural network in this work was optimized using the Bayesian Optimization Algorithm. A heuristic algorithm for real-time crack mapping was introduced and tested on different images with complicated crack position and orientation Also, a video was recorded and processed for testing the real-time ability of the algorithm. Although the database was carefully selected and curated in this work, the authors attempted to include a robust population of crack images to improve the selection and classification power of the CNN. However, this study is limited to only one block size and 13 classification categories. Certainly, for commercial applications, increasing the number of images within the training and increasing the computing power will allow users to reduce the size of the tiles and increase the number of classification categories which may further refine the smoothness of the mapping segments. The mapping results via the CMA may also be used for crack type classification and causation, what type of asphalt is more prone to cracking, what type of asphalts are more suitable for different road conditions with respect to the traffic, how to choose the best asphalt for various conditions and finally estimating the repair and protection costs for each individual road type. Analyzing the crack propagation patterns based on geographical information of the road using the CMA provides more analytical information in combination with other data that could help during the decision making process for road construction.

REFERENCES

1. J. A. Epps and M. Ardila-Coulson, "Summary of SHRP research and economic benefits of pavement maintenance," 1997.
2. National Highway Traffic Safety Administration, "National motor vehicle crash causation survey: Report to congress," Tech. Rep. DOT HS 811 059, National Highway Traffic Safety Administration, 2008.

3. L. Sun, M. Kamaliardakani, and Y. Zhang, "Weighted neighborhood pixels segmentation method for automated detection of cracks on pavement surface images," *Journal of Computing in Civil Engineering*, vol. 30, no. 2, p. 04015021, 2015.
4. A. Ouyang, C. Luo, and C. Zhou, "Surface distresses detection of pavement based on digital image processing," in *International Conference on Computer and Computing Technologies in Agriculture*, pp. 368–375, Springer, 2010.
5. K. Gopalakrishnan, "Advanced Pavement Health Monitoring and Management [WWW Document]," IGI Global Videos, 2016.
6. E. Salari and G. Bao, "Pavement distress detection and classification using feature mapping," in *Electro/Information Technology (EIT), 2010 IEEE International Conference on*, pp. 1–5, IEEE, 2010.
7. G. Sollazzo, K. Wang, G. Bosurgi, and J. Li, "Hybrid procedure for automated detection of cracking with 3D pavement data," *Journal of Computing in Civil Engineering*, vol. 30, no. 6, p. 04016032, 2016.
8. S. Li, Y. Cao, and H. Cai, "Automatic pavement-crack detection and segmentation based on steerable matched filtering and an active contour model," *Journal of Computing in Civil Engineering*, vol. 31, no. 5, p. 04017045, 2017.
9. P. Prasanna et al., "Automated Crack Detection on Concrete Bridges," vol. 13, no. 2, pp. 591–599, 2016.
10. Y. J. Cha, W. Choi, and O. Büyüköztürk, "Deep Learning-Based Crack Damage Detection Using Convolutional Neural Networks," *Computer-Aided Civil and Infrastructure Engineering*, vol. 32, no. 5, pp. 361–378, 2017.
11. C. Koch, K. Georgieva, V. Kasireddy, B. Akinci, and P. Fieguth, "A review on computer vision based defect detection and condition assessment of concrete and asphalt civil infrastructure," *Advanced Engineering Informatics*, vol. 29, no. 2, pp. 196–210, 2015.
12. P. Patidar, M. Gupta, S. Srivastava, and A. K. Nagawat, "Image de-noising by various filters for different noise," *International Journal of Computer Applications*, vol. 9, no. 4, 2010.
13. W. R. Vavrik, L. D. Evans, J. A. Stefanski, and S. Sargand, "PCR Evaluation–Considering Transition from Manual to Semi-Automated Pavement Distress Collection and Analysis," 2013.
14. A. Ayenu-Prah and N. J. E. J. o. A. i. S. P. Attoh-Okine, "Evaluating pavement cracks with bidimensional empirical mode decomposition," vol. 2008, no. 1, p. 861701, 2008.
15. P. Subirats, J. Dumoulin, V. Legeay, and D. Barba, "Automation of pavement surface crack detection using the continuous wavelet transform," in *Image Processing, 2006 IEEE International Conference on*, pp. 3037–3040, IEEE, 2006.
16. K. C. Wang, Q. Li, and W. J. T. R. R. Gong, "Wavelet-based pavement distress image edge detection with a trous algorithm," vol. 2024, no. 1, pp. 73–81, 2007.
17. G. W. Flintsch and K. K. McGhee, *Quality management of pavement condition data collection*. Transportation Research Board, 2009.
18. S. Liang and B. Sun, "Using wavelet technology for pavement crack detection," in *ICLEM 2010: Logistics For Sustained Economic Development: Infrastructure, Information, Integration*, pp. 2479–2484, 2010.
19. Q. Zou, Y. Cao, Q. Li, Q. Mao, and S. J. P. R. L. Wang, "CrackTree: Automatic crack detection from pavement images," vol. 33, no. 3, pp. 227–238, 2012.
20. H. Moon and J. J. P. o. t. t. I. Kim, "Intelligent crack detecting algorithm on the concrete crack image using neural network," pp. 1461–1467, 2011.

21. M. O'Byrne, B. Ghosh, F. Schoefs, and V. J. C. A. C. Pakrashi, "Regionally enhanced multiphase segmentation technique for damaged surfaces," *Infrastructure Engineering*, vol. 29, no. 9, pp. 644–658, 2014.
22. M. O'Byrne, F. Schoefs, B. Ghosh, and V. J. C. A. C. Pakrashi, "Texture analysis based damage detection of ageing infrastructural elements," *Infrastructure Engineering*, vol. 28, no. 3, pp. 162–177, 2013.
23. M. R. Jahanshahi, S. F. Masri, C. W. Padgett, and G. S. J. M. v. Sukhatme, "An innovative methodology for detection and quantification of cracks through incorporation of depth perception," *Machine Vision and Applications*, vol. 24, no. 2, pp. 227–241, 2013.
24. L. Wu, S. Mokhtari, A. Nazef, B. Nam, and H.-B. J. J. o. C. i. C. E. Yun, "Improvement of crack-detection accuracy using a novel crack defragmentation technique in image-based road assessment," vol. 30, no. 1, p. 04014118, 2014.
25. M. S. Kaseko, Z.-P. Lo, and S. G. J. J. o. t. e. Ritchie, "Comparison of traditional and neural classifiers for pavement-crack detection," vol. 120, no. 4, pp. 552–569, 1994.
26. N. Li, X. Hou, X. Yang, and Y. Dong, "Automation recognition of pavement surface distress based on support vector machine," in *Intelligent Networks and Intelligent Systems, 2009. ICINIS'09. Second International Conference on*, pp. 346–349, IEEE, 2009.
27. K. J. D. Gopalakrishnan, "Deep Learning in Data-Driven Pavement Image Analysis and Automated Distress Detection: A Review," vol. 3, no. 3, p. 28, 2018.
28. A. Vedaldi and K. Lenc, "Matconvnet: Convolutional neural networks for MATLAB®," in *Proceedings of the 23rd ACM International Conference on Multimedia*, pp. 689–692, ACM, 2015.
29. K. Gopalakrishnan, S. K. Khaitan, A. Choudhary, and A. J. C. Agrawal, "Deep Convolutional Neural Networks with transfer learning for computer vision-based data-driven pavement distress detection," *Construction and Building Materials*, vol. 157, pp. 322–330, 2017.
30. S. Dorafshan, R. J. Thomas, and M. J. C. Maguire, "Comparison of deep convolutional neural networks and edge detectors for image-based crack detection in concrete," *Construction and Building Materials*, vol. 186, pp. 1031–1045, 2018.
31. H. Maeda, Y. Sekimoto, T. Seto, T. Kashiyama, and H. J. a. p. a. Omata, "Road Damage Detection Using Deep Neural Networks with Images Captured Through a Smartphone," 2018.
32. L. Some, "Automatic image-based road crack detection methods," 2016.
33. Y. Kellouche, B. Boukhatem, M. Ghrici, and A. Tagnit-Hamou, "Exploring the major factors affecting fly-ash concrete carbonation using artificial neural network," *Neural Computing and Applications*, pp. 1–20.
34. G. Xu, J. Ma, F. Liu, and X. Niu, "Automatic recognition of pavement surface crack based on BP neural network," in *Computer and Electrical Engineering, 2008. ICCEE 2008. International Conference on*, pp. 19–22, IEEE, 2008.
35. L. Zhang, F. Yang, Y. D. Zhang, and Y. J. Zhu, "Road crack detection using deep convolutional neural network," in *Image Processing (ICIP), 2016 IEEE International Conference on*, pp. 3708–3712, IEEE, 2016.
36. K. C. Wang and Q. Li, "3D Laser Imaging for ODOT Interstate Network at True1-mm Resolution," Tech. rep., Oklahoma Department of Transportation, 2014.
37. A. Krizhevsky and G. Hinton, "Learning multiple layers of features from tiny images," 2009.
38. I. Goodfellow, Y. Bengio, A. Courville, and Y. Bengio, *Deep learning*. MIT Press Cambridge, 2016.

39. C. M. Bishop, *PATTERN RECOGNITION AND MACHINE LEARNING*. Springer-Verlag New York, 2016.
40. J. Kacprzyk, "Studies in Computational Intelligence, Volume 137," 2008.
41. K. Murphy, "Machine learning: a probabilistic approach," Massachusetts Institute of Technology, pp. 1–21, 2012.
42. R. Baptista and M. J. a. p. a. Poloczek, "Bayesian Optimization of Combinatorial Structures," 2018.
43. M. A. Gelbart, J. Snoek, and R. P. Adams, "Bayesian optimization with unknown constraints," arXiv preprint arXiv:1403.5607, 2014.
44. C. E. Rasmussen, "Gaussian processes in machine learning," in *Advanced lectures on machine learning*, pp. 63–71, Springer, 2004.
45. J. Mockus, *Bayesian approach to global optimization: theory and applications*. Springer Science & Business Media, 2012.
46. E. Brochu, V. M. Cora, and N. De Freitas, "A tutorial on Bayesian optimization of expensive cost functions, with application to active user modeling and hierarchical reinforcement learning," arXiv preprint arXiv:1012.2599, 2010.
47. F. Harrell, "Regression modeling strategies," as implemented in R package 'rms' version 3.3, 2013.
48. Y. Ma, S. Soatto, J. Kosecka, and S. S. Sastry, *An invitation to 3-d vision: from images to geometric models*. Springer Science & Business Media, 2012.
49. S. Qiu, W. Wang, S. Wang, and K. C. J. J. o. C. i. C. E. Wang, "Methodology for Accurate AASHTO PP67-10–Based Cracking Quantification Using 1-mm 3D Pavement Images," vol. 31, no. 2, p. 04016056, 2016.

3 Deep Learning for Efficient Road Damage Detection

Sadra Naddaf-Sh, M-Mahdi Naddaf-Sh, Amir R Kashani, Hassan Zargarzadeh

CHAPTER SUMMARY

Pavement condition evaluation is essential to time the preventative or rehabilitative actions and control distress propagation. Failing to conduct timely evaluations can lead to severe structural and financial loss of the infrastructure and complete reconstructions. Automated computer-aided surveying measures can provide a database of road damage patterns and their locations. This database can be utilized for timely road repairs to gain the minimum cost of maintenance and the asphalt's maximum durability. This chapter introduces a deep learning-based surveying scheme to analyze the image-based distress data in real-time. A database consisting of a diverse population of crack distress types such as longitudinal, transverse, and alligator cracks, photographed using mobile-device is used. Then, a family of efficient and scalable models that are tuned for pavement crack detection is trained, and various augmentation policies are explored. Proposed models, resulted in F1-scores, ranging from 52% to 56%, and average inference time from 178-10 images per second. Finally, the performance of the object detectors are examined, and error analysis is reported against various images. The source code is available at https://github.com/mahdi65/roadDamageDetection2020.

3.1 INTRODUCTION

Road infrastructure and its ability to efficiently and safely handle the transfer of people and goods point to point are vital means in societal, economic growth, and vitality. Long term viability of infrastructure and economy depend on strategies in managing, preserving, and rehabilitating of the road pavements. Moreover, poor pavement conditions are not only a contributing factor in excessive wear on vehicles, but also increase the number of crashes and delays that all lead to additional financial losses [1]. The National Highway Traffic Safety Administration (NHTSA) claims that poor pavement conditions contribute to 16% of traffic crashes [2].

Manual inspection is the dominant technique for pavement distress identification [3]. However, manual inspection can be labor-intensive, costly, and time-consuming. Furthermore, the manual inspection is prone to human visual error, safety issues because of the passing vehicles, and impeding the traffic flow [4].

Performing maintenance operation on millions of miles of pavement is estimated to cost upwards of $25 billion per year [5]. Pavement surveys, including surface

DOI: 10.1201/9781003766575-3

and sub-surface assessments, are needed for effective and efficient maintenance operations. Frequent assessments are used in optimizing and prioritizing preservative and rehabilitative tasks. As current pavements age and new pavements are being added to the network, the manual inspection cannot satisfy timely assessment requirements, and it needs to be enhanced by semi-automated and ultimately automated methods. Moreover, automated pavement distress diagnosis techniques need to be accurate, cost-effective, non-destructive, generalizable, and relatively environmentally and user-friendly [6, 7].

As a solution, deep learning-based methods show exceptional results in pavement distress detection and other applications in recent years [8]. Zhang, et al. [9] utilized deep convolutional neural networks automatic pavement crack detection that was trained with manually annotated image patches acquired by a smartphone. Moreover, multiple approaches and datasets for pavement defect detection are reviewed in [10] including multi-class crack classification [11], and segmentation [12]. By comparing multiple approaches of crack detection problem, Cao, et al. [10] suggested further development in accuracy and real-time performance of the algorithms, robustness, and generalization in various weather conditions.

In a survey in [13], eight models including Faster RCNN Resnet 50, Faster RCNN Resnet 101, Faster RCNN Inception Resnet, Faster RCNN Inception V2, SSD Mobile V1, SSD Mobile V2, SSD Inception V2, and SSDLite Mobile V2 are investigated for pavement crack detection. The authors reported that the highest mAP of 0.54 resulted from the Faster R-CNN Inception-Resnet-V2 model on a randomly selected validation set. Also, the inference time of the same model is estimated 17 seconds for a 600×600 pixel image, and 24.3 seconds for a 3680×2760 pixel image.

In addition, in object detection, there are some common methods like Ensemble Methods[14, 15] and Test-Time Augmentation(TTA) that are used to increase the performance of image recognition. These methods can significantly enhance the final prediction results. For example, [16] fused 16 models in the IEEE BigData Road Damage Detection Challenge 2018 and achieved a mean F1-score of 0.6455 that is the first place in the 2018 challenge. Additionally, top object detection competition winners with high accuracy detection applied similar methods, e.g. [17, 18]. However, the aforementioned method increases inference time, while being real-time is a favorable feature in the pavement crack detection task [10].

There should be a shared database of cracks to justify and compare different method's performances and accuracy. Maeda, et al. [19] created an open-source crack dataset with annotations. This dataset is used for the IEEE BigData Cup Challenges of 2018, and 2020 [20]. This dataset contains annotated images from 3 countries, including India, Japan, and the Czech Republic. Since time performance is one of the challenges in the final crack detection platform [10], it is considered in this chapter as one of the constraints. A family of EfficientDet models is trained and tested for evaluating the accuracy, robustness, and inference latency by utilizing the above road dataset. A series of trained models are prepared and tested utilizing the scalability feature of the selected models. It is possible to select the model with respect to the target hardware without losing accuracy while the performance is maintained. A summery of competition results is provided in [21].

The remainder of this chapter is organized as follows. Section 3.2 gives an overview of the applied method for pavement distress detection. In section 3.3, the data preparation and dataset are explained. The model details, metrics, and experimental results are discussed in the section 3.4. Finally, in the 3.5 the overall results and future improvements are discussed.

3.2 METHOD

In this chapter, to address the trade-offs between accuracy, real-time performance, and scalability in pavement distress detection, a family of one-stage networks called EfficientDet [22] is used. In contrast to one-stage methods, two-stage methods (like Faster-RCNN [23]) can achieve higher accuracy by utilizing region proposal networks. On the other hand, scalability and real-time performance constraints make one-stage and one-scale designs a viable choice, maximizing the accuracy while minimizing the trained model's inference time. The EfficientDet is a single-stage, single-model, and single-scale object detector that can efficiently scale concerning the hardware resource constraints. Moreover, the trained model is deployable on end devices ranging from mobile-device with limited hardware and power resources to multi-GPU workstations.

The first step would be feature extraction from a single image and then using object detection methods to locate cracks in the image. In EfficientDet the multi-scale feature extraction task is performed by EfficientNet backbone [24]. In the pavement distress detection application, features like crack orientation, background, brightness, and the crack area should be taken into consideration. Hence, using multi-scale feature extraction can benefit from the accurate detection task. Feature Pyramid Network (FPN)[25] uses a top-down approach to sum up multi-scale features that can be used for fusing multi-scale extracted features. In the FPN, different scales do not necessarily contribute equally to the output features that can lead to some missing features in the crack detection process.

Bi-directional Feature Pyramid Network (BiFPN) can be used to address the equal contribution issue in FPN. Also, BiFPN exploits trainable weights to learn features with the most contribution in the final model. So, features from layers P3 to P7 from the backbone are passed to BiFPN as the selected multi-scale features. Finally, to obtain the class and box of detected cracks in an image, the output of BiFPN will be fed into box and class prediction networks.

The depth and width of EfficientDet's backbone are following the width and depth scaling coefficients of EfficientNet [24]. The BiFPN and box/class prediction network depth and width in the EfficientDet are also calculated separately. Each model from EfficientDet-D0 to EfficientDet-D7 have a similar architecture to Figure 3.1. It should be noted that from D0 to D7, the size of the backbone, resolution of the input image, the number of repeated BiFPN blocks, and the number of layers in box/class networks increases using a single compound coefficient ϕ. Therefore, using equations (3.1), (3.2), (3.3) input image resolution, BiFPN depth and width, depth of

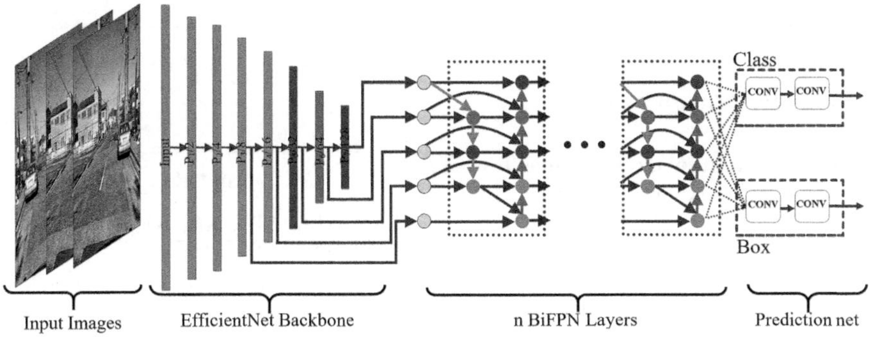

Figure 3.1 EfficientDet family architecture: Images pass through the backbone, and feature scales P3 to P7 get fed into the BiFPN network. Input image resolution is calculated from (3.1). Number of BiFPN layers extracted using (3.2). Depth of box/class prediction net is determined using (3.3).

box/class networks are determined, respectively. Tables 3.2 and 3.3 represent the input image resolution and parameters devised for each model.

$$R_{\text{input}} = 512 + \phi \cdot 128 \tag{3.1}$$

$$W_{\text{bifpn}} = 64 \cdot (1.35^{\phi}), D_{\text{bifpn}} = 3 + \phi \tag{3.2}$$

$$D_{\text{box}} = D_{\text{class}} = 3 + \lfloor \frac{\phi}{3} \rfloor \tag{3.3}$$

3.3 DATASET

In this chapter, the presented dataset in IEEE BigData 2020 Road Damage Detection challenge is used [19]. Images are taken using smartphones mounted on

Table 3.1
Distribution of images and labels.

	Total	Categories			
		D00	D10	D20	D40
Train	18930	5918	4014	7535	5103
Validation	2111	674	432	846	524
Test1	2631	—[a]	—	—	—
Test2	2664	—	—	—	—
Total	26336	6592	4446	8381	5627

[a] Testset annotations were not released.

Table 3.2
Model performance details.

Model Name	Input Image Resolution	Backbone Name	Test1 F1	Test2 F1	Validation						
					AP	AP_{50}	AP_{75}	AP_s	AP_m	AP_l	F1
D0	512	B0	52.1	51.4	19.1	47.2	11.5	7.2	14.3	22.2	54.04
D0-AUG	512	B0	51.2	52.5	19.8	48.4	12.1	7.9	15.4	22.7	54.03
D1	640	B1	53.8	54.7	21.7	51.5	13.4	15.3	16.9	25.0	56.9
D1-AUG	640	B1	54.4	**55.4**	22.0	51.7	13.1	17.1	17.7	24.7	56.5
D2	768	B2	55.2	54.9	22.9	53.5	14.9	10.4	18.6	24.9	56.7
D2-AUG	768	B2	54.1	54.0	22.9	54.2	15.2	13.3	18.8	24.7	56.6
D3	896	B3	**56.5**	54.7	23.0	53.4	15.0	10.5	18.4	25.4	56.5
D3-AUG	896	B3	56.3	54.2	22.6	53.4	14.7	11.4	18.3	24.8	56.8
D4	1024	B4	54.53	54.6	22.8	53.3	15.1	15.5	18.1	25.7	57.2
D7-AUG	1536	B6	**56.5**	**54.9**	23.4	53.6	15.0	31.4	19.2	25.5	56.5

Table 3.3
Inference Time and other parameters related to each model.

Model Name	#Params	Batch size/ Learning rate	Inference Time(img/s) V100		
			b^b=1	b^b=8	b^b=16
D0	3.8M	90/0.112	20	121	178
D1	6.5M	75/0.075	15	98	147
D2	8M	45/0.056	12	82	100
D3	11.9M	18/0.026	10	54	58
D4	20.5M	9/0.011	9	35	37
D7	51M	8/0.01	6	9	10

b denotes batch size for inference.

the dashboard vehicles. Annotations for this dataset are provided in PASCALVOC format[26]. Crack types are named based on the Japan Road Association(JRA)[27] standards. The cracked area is annotated using a rectangular box, and each box is associated with a crack class label. In the 2020 challenge, four classes of interest consist of Longitudinal crack(D00), Transverse crack(D10), Alligator Crack(D20), and Potholes(D40).

The distribution of classes in the dataset is presented in Table 3.1. A total of 26336 images are available in the dataset. Challenge organizers withheld 20% of images (i.e., 5295) as test a set and released the rest of the images (i.e., 21041) with annotations to the contestants. The test set is divided into two sections; Test1 includes 2631 images, and Test2 includes 2664 images. The test set annotations are not released to the participants. The evaluation of the test set is performed online and with a limited number of tries per day, and the feedback is F1-score.

3.4 EXPERIMENTS

In this section, firstly, metrics including F1-score and Average Precision(AP) for analyzing and evaluating the results, are discussed. Then, network training, augmentation transferring, and hyperparameter tuning for pavement crack detection are studied. Finally, in a discussion, multiple models trained with various parameters and their contribution to the accuracy, time performance, and F1-score are compared.

3.4.1 EVALUATING METRICS

For a fair comparison of experiment results, two sets of metrics are applied. First, measuring the F1-Score due to competition rules and also to submit/compare results with other participating teams. Second, Average Precision is used to be able to measure precision-recall on different confidence thresholds.

F1-Score: Competition Defined metrics as used in [20] are as the following:

$$F1 = 2 \cdot \frac{precision \cdot recall}{precision + recall} \tag{3.4}$$

where,

$$precision = \frac{tp}{tp + fp}; recall = \frac{tp}{tp + fn} \tag{3.5}$$

in each of which tp is equal to the number of true positive detections of the model, fp is the number of objects that model incorrectly marked as crack, and fn indicates the number of existing objects(cracks) in the image that predictor has not detected. A bounding box is considered as correct when the area of the ground truth bounding box and the detected box have at least 0.5(based on competition rule) Intersection over the area of the union of two boxes, which is called Intersection over Union(IoU). Also, the class labels of both bounding boxes should be the same, that means:

$$IoU = \frac{area(B_{\mathrm{Bp}} \cap B_{\mathrm{gt}})}{area(B_{\mathrm{Bp}} \cup B_{\mathrm{gt}})}. \tag{3.6}$$

Average Precision (AP): is also used to demonstrate and compare results. F1-score results are reported using only one confidence level. However, recall and precision of a robust object detector are not altering much, with varying confidence. The area under curve for precision-recall curve will not get precise results due to zig-zag like curve[29]. Instead, defining all-point interpolation can obtain accurate results by pruning zig-zag behavior of the precision-recall curve. AP has become a standard for comparing model performance in different object detection challenges[30, 28, 26, 31] as well as in literature[32, 22, 23, 33]. In section 3.4.5, models are evaluated using mAP(mean AP)(where with setting IOU threshold from 0.50 to 0.95 with 0.05 step mAP is calculated), AP_{50}, AP_{75}, AP_s(s stands for small and objects with area $< 32^2$) , AP_m(m is medium and area of the objects are between 32^2 and 96^2), as well as AP_l(objects with area $> 96^2$).

3.4.2 ANCHOR BOXES

EfficientDet box prediction network utilizes anchor boxes to enhance the detection of overlapping bounding boxes. The default value for the box prediction network consists of three aspect ratios (i.e., 0.5, 1.0, 2.0). For the crack detection task, based on all labeled bounding boxes in the dataset, the default value is increased to seven (and in larger models eight) aspect ratios. Also, k-means clustering [1] is used to find the set of optimal aspect ratios for box prediction network [32]. Moreover, the input image resolution is also considered in aspect ratio calculation. Other parameters left as it is in the code repository of [22].

3.4.3 TRAINING AND HYPERPARAMETERS TUNING

For training the EfficientDet D0-D4 models, a GPU cluster with three V100-16GB NVIDIA GPUs is used. The input image size in EfficientDet D7 is 2.5x of the original image size. Hence, more computational power is needed, and the model is trained on a GPU cluster with eight V100-32GB NVIDIA GPUs. Also, a technique for fitting large models in limited GPU memory is to employ Gradient Checkpointing [34], even though training time increases significantly. Batch size is proportional to the model size due to GPU memory limitation. This limitation is more observable in D3-D7 models than D0-D2. In distributed training, synchronized cross-GPU batch normalization(syncBN) is applied, which provides access to cross-device batch normalization to improve statistics.

During training, before each image is fed to the network, it is first randomly flipped horizontally and/or resized and finally normalized using mean and standard deviation values that are over each RGB channel. Also, for all models, mixed-precision training is used. Mixed precision[35] utilizes performing half-precision operations, leading to faster inference and training and decreased memory usage. In the training process, decreasing memory usage makes it possible to fit larger models and higher-resolution images to the GPUs.[2] Moreover, pre-trained weights on MS COCO [22] that are converted to Pytorch [36] are utilized as initial weights. Furthermore, all layers are trained, and the layers' weights are not frozen.

The learning rate and batch sizes are listed in Table 3.3. Also, the cosine learning rate is used [37]. In the beginning of training for the first few epochs (3-5) learning rate increases gradually to the desired point, and from epoch 5 to the end of

---

[1] partially used https://github.com/mnslarcher/kmeans-anchors-ratios

[2] apex package for mixed-precision training available at https://nvidia.github.io/apex

the training process learning rate decreases gradually in a cosine form. In addition, learning rate noises applied to 30% and 90% of the training process. Moreover, an exponential moving average with weight decay of 0.9998 is applied to stabilize training, especially on larger models that are trained with relatively small batch size. Non-Maximum Suppression(NMS) is applied to prune overlapping bounding boxes through inference. In addition, based on validation results, the best threshold per class is extracted and applied during inference. Table 3.3 illustrates the inference latency and model size for each model.

3.4.4 TRANSFERRING AUGMENTATION POLICIES

Data augmentation not only improves accuracy but also generalizes the trained model [38]. There are countless possibilities for augmentation strategy, including geometric transforms, brightness and color balance adjustment, histogram equalization, and rotation. It is possible to use multiple augmentation strategies on a single image with a varying range of magnitudes. However, selecting optimal augmentation strategies that improve the results and enhance performance is a time-consuming task. Some techniques are developed, such as AutoAugment [39], and RandAungmet [40] to address this issue, which can extract the best set of policies through reinforcement learning. For finding the optimal policy, the model should be trained and evaluated with each set of proposed policies. There are four default policies introduced in [38], which are called policyV0-policyV3. One of the shared augmentation strategies in policyV0 to policyV3 is rotation. By using the rotation strategy, crack orientation will change, i.e., it converts a Longitudinal crack(D00) to Transverse crack(D10) or conversely. Therefore, the policies V0-V3 are updated by eliminating this strategy.

A D0 model with randomly selected 25% of the training set is trained for 150 epochs and evaluated on the validation set with all V0-V3 policies. For comparison, the same process is repeated without applying any augmentation. As it is depicted in Table 3.4, policies V1 and V2 increased AP metrics more than others. Since all AP metrics are increased using policy V1, it is selected as the augmentation policy on the entire dataset during training. It should be noted that the augmentation strategies introduced in [38] use bounding box-only augmentation. In this chapter, the same augmentation strategies are applied to the entire image.

3.4.5 EVALUATING RESULTS

The best F1-scores on both test sets are achieved through model D7, referring to Table 3.2. Some sample results from model D7 are depicted in Figs. 3.2, 3.3, and Fig. 3.4. From analyzing the results in multiple samples, it can be mentioned:

- False Positive (FP) detections: There are three major contributing factors in FP detection. First, the model detects too close or overlapped bounding boxes with the same labels as a single joint bounding box, while in the ground truth the close cracks are annotated with multi bounding boxes. Although the crack is correctly detected, the IoU between the detected joint

Table 3.4
Policies used to train on 25% of train images.

policy Name	Base (% of Improvement)			
	AP_{50}	AP_s	AP_m	AP_l
NoAugment	31.7	3.8	7.3	12.7
policyV0[a]	31.9(**+0.6**)	7.0(**+84.2**)	7.8(**+6.8**)	11.9(**-6.2**)
policyV1[a]	34.1(**+7.5**)	5.9(**+55.2**)	8.2(**+12.3**)	12.9(**+1.5**)
policyV2[a]	33.0(**+4.1**)	6.4(**+68.4**)	8.6(**+17.8**)	12.2(**-4**)
policyV3[a]	33.4(**+5.3**)	5.1(**+34.2**)	7.7(**+5.4**)	13.4(**+5.5**)

[a] Rotation strategies are removed from these policies.

bounding box and ground truths falls behind the desired IoU threshold(i.e., 0.5). Hence, this mismatch between bounding boxes increases both FP and FN. For instance, Fig.3.4b depicts two boxes with Alligator Crack(D20) annotated closely as ground truths, but the model detects the entire region as D20 type. Second, in some cases, patched surface edges(or construction joint part) are detected as a crack (commonly mistaken with Longitudinal Crack(D00) or Transverse Crack(D10) classes). As depicted in Fig. 3.4a, edges of the patched section of the pavement are detected as Transverse Crack(D10) by error. Third, manholes with area < 800 pixels can be detected as potholes(D40).

- False Negative(FN) detections: The contributing factors in FN detection can be counted as follows. First, a bounding box with a small area (< 600), as it is shown in Fig. 3.4c. Second, different background color, e.g., a specific area or pavement markings on the road or the surface of the road is covered with debris3.3c.
- Misclassifications: In some cases, diagonal crack misclassification has occurred between classes D00 and D10.

Some suggested remedies for improving the results can be mentioned as follows. Based on the crack detection results, setting up some ground rules for annotating the dataset like merging close bounding boxes can decrease FP rates. Moreover, expanding the dataset with more image samples of patched pavement surfaces and manholes can help to decrease the FP rate. Also, mounting the camera in an optimal position which covers more area of the pavement would result in more practicable samples and accurate detection.

In addition, it is shown that augmentation could improve results. Referring to Table 3.4, the results of augmentation on AP_{50}, AP_s, and AP_m across all policies shows an improvement. Although the AP_l results on the same table, is improved in

Figure 3.2 Examples of Network Prediction in TX: Photos are taken in South East Texas roads with smartphone showing that network is able to predict damages accurately in unseen areas.

some cases, it is observed that policies like V0 and V2 can decrease the accuracy on AP_l. Many parameters could contribute to this decremental behavior, like the distribution of labels on the large boxes. Regardless of this behavior, the average

(a) (b) (c)

Figure 3.3 Examples of Network Prediction: True Positives 3.3a and 3.3b (green boxes denote ground truth and others are network prediction) and False Negatives 3.3c, where the background of damage is not gray.

(a) (b) (c)

Figure 3.4 Examples of Network Prediction. False Positive of construction joint part-3.4a, False Positive of two close bounding boxes as one crack-3.4b, and False Negatives on tiny bounding boxes(D40)-3.4c.

improvement across all AP metrics, especially AP_{50} and AP_s can be a viable decision factor for selecting an optimal policy.

As an instance, policies V0 and V2 are neglected because of the inclines in AP_l. In contrast, policies V1 and V3 both have improved all APs. So, the policy V1 is a viable option for augmentation, because it improves AP_{50}, AP_s, and AP_m further than policy V3. During training models using augmentation, it is observed that the selected policy V1 can effectively boost AP_s in smaller models D0 and D1. On the other hand, V1 for larger models, i.e., D2-D7, had less improvement over AP_s. Nonetheless, utilizing augmentation does not change model F1-score for larger models.

As the goal in this chapter is set to achieve a balance between time performance and accuracy, referring to Tables 3.2 and 3.3 makes it possible to choose between the D0-D7 models based on the available computational resources. For instance, for smaller embedded devices with limited power, D0 is more applicable for real-time purposes, while the accuracy is declined by 3%. On the other hand, in the case of using a workstation with multiple GPUs, deeper models could address the real-time constraint. Inference Time is obtained by considering the post-processing of input images. Timings with batch size 1 can be improved with further optimization of post-processing overhead.

To test the robustness and generalizability of the model, as it is mentioned in [19], the model is tested on a set of crack images that are photographed in Southeast Texas, USA. Although the model is trained based on images from Japan, Czech, and India, it is possible to use the same model on the other regions as well. In Fig. 3.2, four image samples from the USA are presented.

3.5 CONCLUSIONS

In this chapter, a deep-learning approach is provided to train scalable and efficient models to detect road defects, and also, the model reached competitive accuracy with 56% F1-score. By considering Inference time, the code is able to predict in real-time using mobile devices. Although one can use TTA and ensemble learning to fuse the prediction results of the network, the inference time will dramatically increase. Moreover, transferred augmentation policies enhanced small networks by 2% on F1-score.

For future work, it is suggested to evaluate the reliability of the network utilizing a test set from a new country. Also to apply gradient check pointing to reduce memory cost during training and searching for optimal augmentation policies with machine learning methods and implementing bounding box the only augmentation can be studied further. Finally, to expand and clean the dataset as well as installing a camera with the optimal orientation that covers more of the pavement.

REFERENCES

1. L. Sun, M. Kamaliardakani, and Y. Zhang, "Weighted neighborhood pixels segmentation method for automated detection of cracks on pavement surface images," *Journal of Computing in Civil Engineering*, vol. 30, no. 2, p. 04015021, 2016.
2. National Highway Traffic Safety Administration, "National motor vehicle crash causation survey: Report to congress," National Highway Traffic Safety Administration Technical Report DOT HS 811 059, 2008.
3. A. Ouyang, C. Luo, and C. Zhou, "Surface distresses detection of pavement based on digital image processing," in *International Conference on Computer and Computing Technologies in Agriculture*. Springer, 2010, pp. 368–375.
4. K. Gopalakrishnan, "Advanced pavement health monitoring and management," *IGI Global Videos*, 2016.
5. J. A. Epps and M. Ardila-Coulson, "Summary of SHRP research and economic benefits of pavement maintenance," Federal Highway Administration, Tech. Rep., 1997.
6. G. Sollazzo, K. Wang, G. Bosurgi, and J. Li, "Hybrid procedure for automated detection of cracking with 3D pavement data," *Journal of Computing in Civil Engineering*, vol. 30, no. 6, p. 04016032, 2016.
7. M. M. Dargahi and D. Lattanzi, "Spatial statistical methods for complexity-based point cloud analysis," in *ASME 2020 Conference on Smart Materials, Adaptive Structures and Intelligent Systems*. American Society of Mechanical Engineers Digital Collection, 2020.

8. M. Naddaf-Sh, H. Myler, H. Zargarzadeh et al., "Design and implementation of an assistive real-time red lionfish detection system for AUV/ROVs," *Complexity*, vol. 2018, 2018.
9. L. Zhang, F. Yang, Y. D. Zhang, and Y. J. Zhu, "Road crack detection using deep convolutional neural network," in *2016 IEEE International Conference on Image Processing (ICIP)*. IEEE, 2016, pp. 3708–3712.
10. W. Cao, Q. Liu, and Z. He, "Review of pavement defect detection methods," *IEEE Access*, vol. 8, pp. 14531–14544, 2020.
11. M. Naddaf-Sh, S. Hosseini, J. Zhang, N. A. Brake, and H. Zargarzadeh, "Real-time road crack mapping using an optimized convolutional neural network," *Complexity*, vol. 2019, 2019.
12. M. D. Jenkins, T. A. Carr, M. I. Iglesias, T. Buggy, and G. Morison, "A deep convolutional neural network for semantic pixel-wise segmentation of road and pavement surface cracks," in *2018 26th European Signal Processing Conference (EUSIPCO)*. IEEE, 2018, pp. 2120–2124.
13. M.-T. Cao, Q.-V. Tran, N.-M. Nguyen, and K.-T. Chang, "Survey on performance of deep learning models for detecting road damages using multiple dashcam image resources," *Advanced Engineering Informatics*, vol. 46, p. 101182, 2020.
14. R. Solovyev, W. Wang, and T. Gabruseva, "Weighted boxes fusion: ensembling boxes for object detection models," *arXiv preprint arXiv:1910.13302*, 2019.
15. O. Okun, G. Valentini, and M. Re, *Ensembles in Machine Learning Applications*, vol. 373. Springer Science & Business Media, 2011.
16. Y. J. Wang, M. Ding, S. Kan, S. Zhang, and C. Lu, "Deep proposal and detection networks for road damage detection and classification," in *2018 IEEE International Conference on Big Data (Big Data)*. IEEE, 2018, pp. 5224–5227.
17. T. Gabruseva, D. Poplavskiy, and A. Kalinin, "Deep learning for automatic pneumonia detection," in *Proceedings of the IEEE/CVF Conference on Computer Vision and Pattern Recognition Workshops*, 2020, pp. 350–351.
18. "Statoil/C-CORE Iceberg Classifier Challenge," 2018. [Online]. Available: https://www.kaggle.com/c/statoil-iceberg-classifier-challenge/discussion/48241
19. H. Maeda, Y. Sekimoto, T. Seto, T. Kashiyama, and H. Omata, "Road damage detection using deep neural networks with images captured through a smartphone," *arXiv preprint arXiv:1801.09454*, 2018.
20. D. Arya, H. Maeda, S. K. Ghosh, D. Toshniwal, A. Mraz, T. Kashiyama, and Y. Sekimoto, "Transfer learning-based road damage detection for multiple countries," *arXiv preprint arXiv:2008.13101*, 2020.
21. D. Arya, H. Maeda, S. K. Ghosh, D. Toshniwal, H. Omata, T. Kashiyama, and Y. Sekimoto, "Global road damage detection: State-of-the-art solutions," *arXiv preprint arXiv:2011.08740*, 2020.
22. M. Tan, R. Pang, and Q. V. Le, "Efficientdet: Scalable and efficient object detection," in *Proceedings of the IEEE/CVF Conference on Computer Vision and Pattern Recognition*, 2020, pp. 10781–10790.
23. S. Ren, K. He, R. Girshick, and J. Sun, "Faster r-cnn: Towards real-time object detection with region proposal networks," in *Advances in Neural Information Processing Systems*, 2015, pp. 91–99.

24. M. Tan and Q. V. Le, "Efficientnet: Rethinking model scaling for convolutional neural networks," *arXiv preprint arXiv:1905.11946*, 2019.
25. T.-Y. Lin, P. Dollár, R. Girshick, K. He, B. Hariharan, and S. Belongie, "Feature pyramid networks for object detection," in *Proceedings of the IEEE Conference on Computer Vision and Pattern Recognition*, 2017, pp. 2117–2125.
26. M. Everingham and J. Winn, "The pascal visual object classes challenge 2012 (voc2012) development kit," *Pattern Analysis, Statistical Modelling and Computational Learning, Tech. Rep*, vol. 8, 2011.
27. "Maintenance and repair guide book of the pavement," Japan Road Association, Tokyo, Japan, 2013.
28. T.-Y. Lin, M. Maire, S. Belongie, J. Hays, P. Perona, D. Ramanan, P. Dollár, and C. L. Zitnick, "Microsoft coco: Common objects in context," in *European Conference on Computer Vision*. Springer, 2014, pp. 740–755.
29. R. Padilla, S. L. Netto, and E. A. B. da Silva, "A survey on performance metrics for object-detection algorithms," in *2020 International Conference on Systems, Signals and Image Processing (IWSSIP)*, 2020, pp. 237–242.
30. "COCO Detection Challenge (Bounding Box)," 2020. [Online]. Available: https://competitions.codalab.org/competitions/20794
31. M. Everingham, L. Van Gool, C. K. Williams, J. Winn, and A. Zisserman, "The pascal visual object classes (voc) challenge," *International Journal of Computer Vision*, vol. 88, no. 2, pp. 303–338, 2010.
32. J. Redmon and A. Farhadi, "YOLO9000: better, faster, stronger," in *Proceedings of the IEEE Conference on Computer Vision and Pattern Recognition*, 2017, pp. 7263–7271.
33. T.-Y. Lin, P. Goyal, R. Girshick, K. He, and P. Dollár, "Focal loss for dense object detection," in *Proceedings of the IEEE International Conference on Computer Vision*, 2017, pp. 2980–2988.
34. T. Chen, B. Xu, C. Zhang, and C. Guestrin, "Training deep nets with sublinear memory cost," *arXiv preprint arXiv:1604.06174*, 2016.
35. P. Micikevicius, S. Narang, J. Alben, G. Diamos, E. Elsen, D. Garcia, B. Ginsburg, M. Houston, O. Kuchaiev, G. Venkatesh, and H. Wu, "Mixed precision training," in *International Conference on Learning Representations*, 2018.
36. "EfficientDet (A PyTorch implementation of EfficientDet)," 2020. [Online]. Available: https://github.com/rwightman/efficientdet-pytorch
37. I. Loshchilov and F. Hutter, "Sgdr: Stochastic gradient descent with warm restarts," *arXiv preprint arXiv:1608.03983*, 2016.
38. B. Zoph, E. D. Cubuk, G. Ghiasi, T.-Y. Lin, J. Shlens, and Q. V. Le, "Learning data augmentation strategies for object detection," *arXiv preprint arXiv:1906.11172*, 2019.
39. E. D. Cubuk, B. Zoph, D. Mane, V. Vasudevan, and Q. V. Le, "Autoaugment: Learning augmentation strategies from data," in *Proceedings of the IEEE Conference on Computer Vision and Pattern Recognition*, 2019, pp. 113–123.
40. E. D. Cubuk, B. Zoph, J. Shlens, and Q. V. Le, "Randaugment: Practical automated data augmentation with a reduced search space," in *Proceedings of the IEEE/CVF Conference on Computer Vision and Pattern Recognition Workshops*, 2020, pp. 702–703.

Section II

Digital X-Radiography

4 AI-Powered Weld Assessment with Digital Radiography

M-Mahdi Naddaf-Sh, Sadra Naddaf-Sh, Hassan Zargarzadeh, Sayyed M Zahiri, Maxim Dalton, Gabriel Elpers, Amir R Kashani

CHAPTER SUMMARY

Continuous monitoring and advance inspection are key parts of modern manufacturing and sustainment of aging infrastructure. The growing demand for these needs and shortage of required skill set can slow down the global economy. The diversity of requirements and specialized standards and codes around the world along with time-sensitive aspect of such inspection makes it a prime application for utilizing Artificial Intelligence (AI) as an assisting tool. Welding as a method of jointing metals goes back to Bronze age and to this date is one of the most trusted methods of manufacturing heavy encampments and metalic infrastructures. Non-destructive testing (NDT) is commonly used to verify the quality of welding process or define the need to rework, because of the sensitivity of joints involved and their impact on the reliability of assets. In most critical cases, NDT must be done once right after the weld is being done and then on a scheduled or unscheduled basis as the weld goes through aging. One of most reliable NDT methods is Radiography imaging which is done using penetrating gamma or X-radiation.

Current Assisting Defect Recognition (ADR) tools in the literature are heavily focused on high-quality x-ray images which in many cases are not available in reality. Moreover, the classification of defect is critical. In this chapter and demonstration we review the application of deep learning to address the While available X-ray image's database of welds are very limited, outdated and not close enough to realistic applications in the industry, many suggested methods in articles are not able to be used in real-world application.

In this work first we developed and contextualized a 100000+ number of X-ray images from various welds and annotated them with a group of inspection experts to collect all the related features and utilize those contexts. Moreover, an optimized Convolutional Neural Network (CNN) are designed and trained for detecting defects. Performance of the designed CNN is tested against other CNN architectures and accuracy of 96% is achieved.

DOI: 10.1201/9781003766575-4

4.1 INTRODUCTION

Welded joints are dispensable for industries such as marine, chemical, and aeronautical industries [1]. Digitizing and automation of monitoring the quality of product using AI is one of the main pillar of industry 4.0. NDT is one of the oldest techniques to do this, which is used predominantly in welded joints inspection. Utilization of NDT techniques is inevitable to certify the safety, and reliability of these structures. One of the NDT methods is radiography or X-ray image testing, which is common among industries. However, the process of X-ray images is both time-consuming and subjective for the expert operators [2]. The advent of novel image processing techniques and pattern recognition has applied to accelerate the process of weld defect diagnosis and increase the accuracy of defect detection. Nonetheless, due to the complexity of the task, many efforts have been made to develop such an automated intelligent system to recognize defective welded joints.

Previous research on weld defect detection mostly performed using techniques such segmentation, and texture features extraction and classification, one of which has reached the highest accuracy of 90.91% by D Mery [3]. Artificial Neural Networks (ANNs) is used prevalently with enhancements on classifier description among researches. Kumar et al. [4] used gray-level co-occurrence (GLCM) texture features as ANN input and reached an accuracy of 86.1% for defect classification.

In another work[5], they added geometrical features and enhanced the accuracy to 87.34%. Juan Zapata [6] utilizes two neural classifiers, adaptive network-based fuzzy inference system (ANFIS) and ANN. In their work, ANN achieved accuracy of 78.9%, and using the 12 chosen geometrical features extracted from X-ray images as input, ANFIS reached an accuracy of 82.6%. To both integrate texture and geometrical features and only select useful features preventing an increase in computational complexity, Valavanis et al. [7] employed the sequential backward selection (SBS) technique to design classifier of the ANN, which resulted into 85.4% accuracy.

Wang et al. [8] designed a weld flaw detection system, in which they extracted 12 different features from radiography images such as size, intensity, direction, location, and shape for the input of networks. They applied two well-known networks for detection of weld flaws: fuzzy k-nearest neighbor (fuzzy K-NN) and multi-layer perceptron (MLP) neural networks, which attained accuracy of 91.57% and 92.39%, respectively. However, traditional feature extraction methods used in these systems are not able to extract discriminant features.

The solution to improve pattern recognition in weld flaw detection is tacking advantage of deep neural networks that can extract higher and more profound level features. Jin et al. [9] employed Deep Convolutional Neural Networks (DCNN) on three resampled datasets based on GDXray [10] weld dataset and the highest achieved accuracy is 97.2% on the dataset which synthetic minority over-sampling technique (SMOTE) were performed on. For the remaining datasets two techniques of random over-sampling (ROS) and random under-sampling (RUS) for modifying samples were applied, in that best case performances were 96.3% and 79.9% for each methods, respectively. Integration of computer vision methods using DCNNs shows exceptional promise for use in NDT applications, but require many images

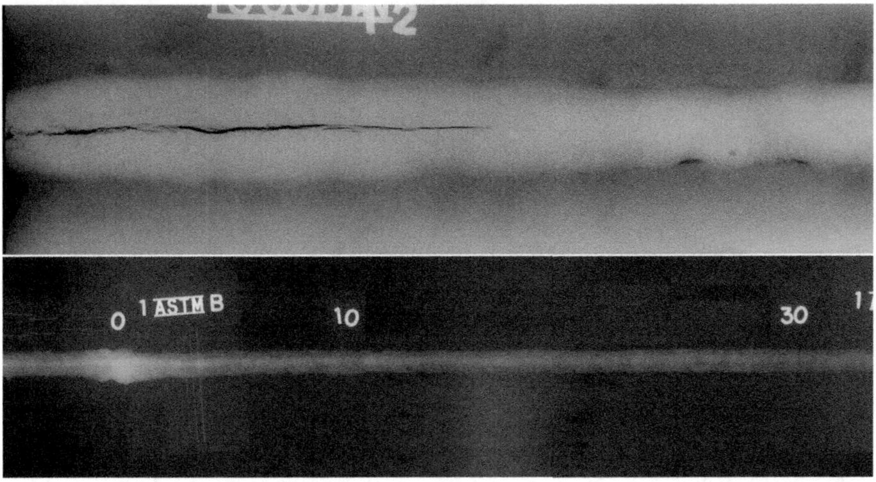

Figure 4.1 Tow samples of images in GDXray database (Top) and SBD database (Bottom). The bottom image is cropped to be able to compare with the other. Defects are more visible in the GDXray database.

for the training process. Although utilizing CNNs improves the defect detection accuracy, there are known drawbacks in the conventional approaches for studying the defects.

As reviewed above, deep learning-based methods for NDT improves false detection accuracy within x-ray images, whether using different CNN architectures or pre-trained models. In the reported works databases were based on the GDXray database which is public [10], and WDXI dataset [11] which is not available publicly. As depicted in Figure 4.1 (Top image), defects in images of the GDXray database are handpicked and defects are easily visible due to their extreme nature. Also, weld images in the GDXray database are limited to 88 samples. Although, GDXray is become a standard benchmark for testing different algorithms performances [9] but under the real world condition defects in the x-ray images are not perfectly noticeable like that in the GDXray. As it is shown in Figure 4.1 (Bottom image), since defects in x-ray images are not handpicked and excessive (and more realistic to working condition of welding operations) it is more challenging to detect defects. Moreover, in previous studies whether traditional image processing algorithms or deep learning based methods were used, difference in wall thickness of welded objects and scattering or waving in the weld root are also neglected. In such cases due to change of thickness, intensity of image varies significantly in each segment of the weld and surface as it is shown in Figure 4.2.

Regarding the nature of images in SBD database, in this chapter the main goal is set to locate and identify defects (detection) and to determining the defect type. Based on the mentioned differences with GDXray database the proposed

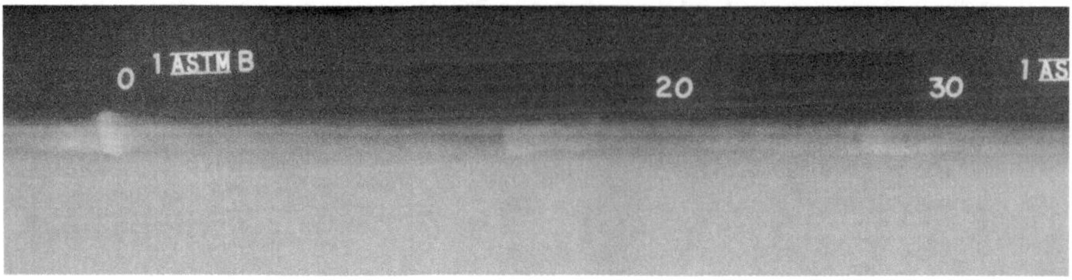

Figure 4.2 An image sample from SBD database in that two surfaces with different thicknesses are welded.

methods either not detecting the defects in SBD dataset at all and the overall accuracy is decreased dramatically. In order to address this problem a deep convolutional neural network is designed and trained to be able to detect defects in SBD database.

4.2 DATABASE PREPARATION

To develop and evaluate proposed network architecture we used full size images containing defects, 15360 × 1024 (*Width* × *Height*) pixels, of a welded pipeline were selected. All full sized images were cropped base on the weld's center in each image. The patches were 224 × 224 pixels. Total of 100000 patches were categories as non-defected and defected by an expert with API and ASME welding certificates. And we selected a smaller set of non-defected and defected patches in a balanced number at the first phase, and the database is named SBD-1. In addition, for evaluating network's performance against different types of defects, using a new set of 23371 defective patches, another database SBD-2 was created that contains 11 types of defects. Defect types and quantity are available in Table 4.1. Figure 4.2, depicts 10 samples of non-defected patches, and in Figure 4.3, 11 patches along with defect type are shown.

4.3 EXPERIMENTAL RESULTS

Several convolutional neural network architectures like VGG-16, VGG-19, AlexNet and ResNet are applied on SBD-1. The maximum accuracy resulted with these well-known architectures is 86%. As depicted in Figure 4.5, an optimized CNN architecture for weld defect detection is proposed in this article. Figure 4.5 illustrates the overall CNN architecture-input and multiple convolutional layer, followed by a batch normalization layer, ReLU layer, max-pooling layer, fully connected (FC) layer, SoftMax layer, and output layer for the classification task. The first layer is the input layer that receives the input patch for classification. The defect features in each batch were extracted using multiple convolution layers. This layer consists of various sets of neurons whose weights and biases will be updated relative to the defect

Table 4.1
Defect Types

Defect Type	Total Number of Patches
Elongated Slag Inclusions (ESI)	5884
Hallow Bead (HB)	5204
Isolated Slag Inclusions (ISI)	3776
Gas Porosity (GP)	2528
Inadequate Penetration (IP)	1216
External Undercut (EU)	692
Porosity (P)	785
Scattered Porosity (SP)	632
Inadequate Penetration Due to high-low (IPD)	938
Internal Undercut (IU)	902
Internal Concavity (IC)	814

features. In the convolutional layer, the neuron input consists of small sectors from the previous layer that is called the filter (kernel). The size of the filter, can be tuned from 1 × 1 pixels up to the size of the input image. In the convolution layer, the filter moves along the input and builds a convoluted feature map. To increase the number of feature maps, multiple filters should be used, and each filter has different weights and biases to be able to extract various features of the image. The stride (amount of horizontal and vertical movement of the filter on the input per convolution) is set to 3 pixels.

After the convolution layer, a batch normalization layer is used for reducing the CNN sensitivity to initial hyperparameter values and decreasing training processing time. Following the batch normalization layer, a Rectified Linear Unit (ReLU) activation layer is added to apply a zero threshold to all negative values in the batch

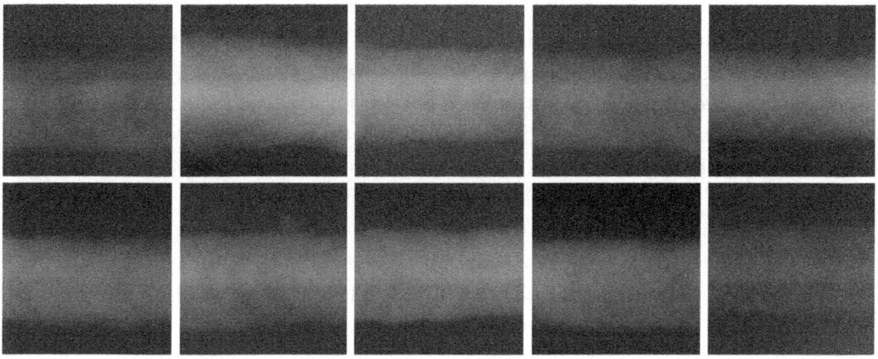

Figure 4.3 10 samples of non-defect images.

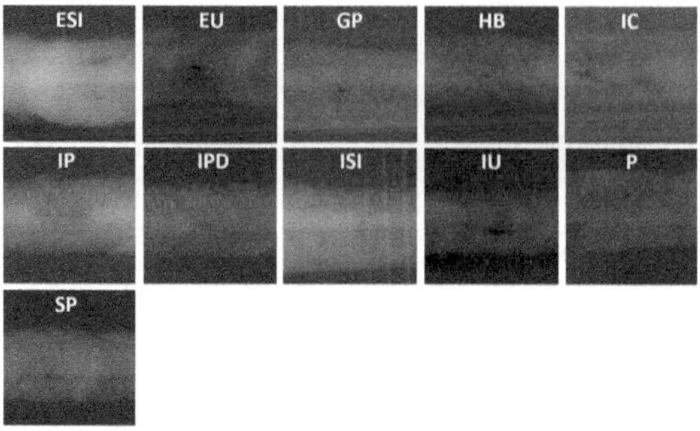

Figure 4.4 Classes of Welding Defects based on ASME standard.

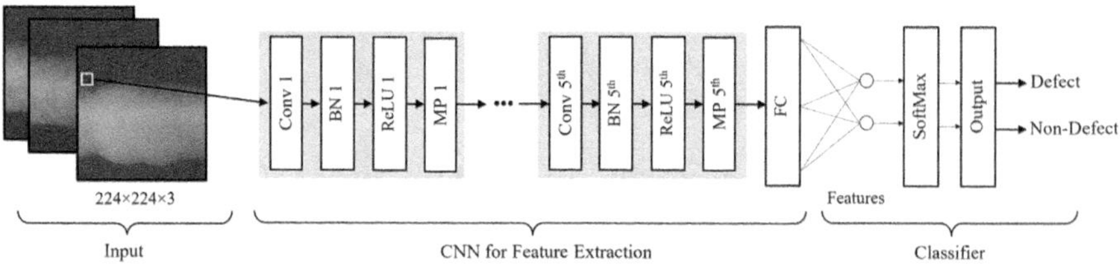

Figure 4.5 Network architecture used for weld defect detection. 5 convolution layers are used in this architecture. Conv: convolution layer, BN: batch normalization layer, MP: max-pooling layer, and FC: fully connected layer.

normalization layer. The max-pooling layer down-samples the input by dividing it into rectangular pooling regions to compute the maximum of each region of gathered feature matrices. After designing the feature extractor, the fully connected (FC) layer is used to map the features matrix in the last layer in the form of a vector, where is chosen equal to the number of categories in the database which in this case are 2. Stochastic Gradient Descent with Momentum (SGDM) was used to train the CNN for classification. This method updates CNN's weights and biases to minimize the loss function that measures the difference between true classified and false classified patches. Moreover, to obtain optimized values for number of layers, filter size and other hyperparameters in the proposed architecture, the Bayesian optimization algorithm is used as described in [12]. By applying this architecture on the first database detection accuracy is increased to 96%.

For the SBD-2 database both trained optimized and un-optimized CNN architectures are tested. Among un-optimized CNN architectures the VGG-19 has the highest accuracy. In contrast the optimized CNN architecture (as shown in Figure 4.5), is able to increase the accuracy up to 10%. The accuracy improvement by using the proposed optimized CNN can be found in Figure 4.6.

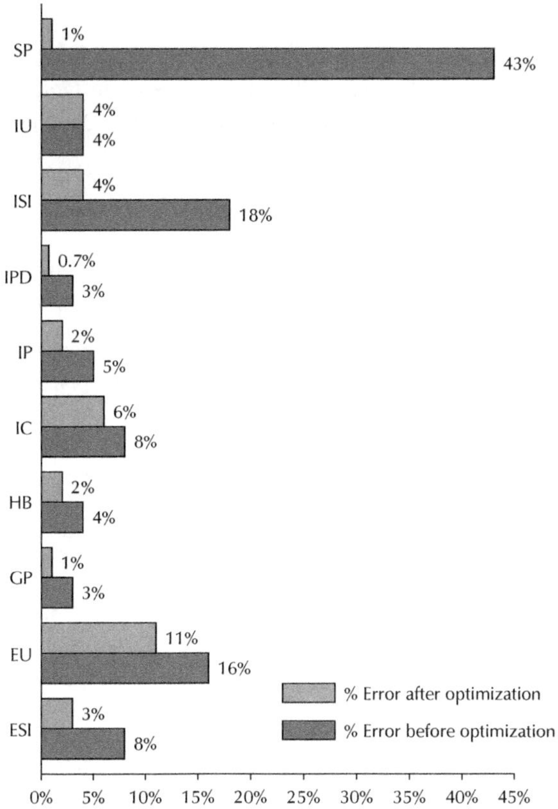

Figure 4.6 Comparison between error among 10 defect categories before and after optimization.

4.4 CONCLUSION

In this work two realistic welding quality datasets for training deep learning models were created based on radiography images collected from various projects and NDT experts annotation SBD-1 and SBD-2. An optimized CNN is designed in order to find defects in the weldment and heat affected zones, and then it is trained and evaluated based on prepared datasets. An accuracy of 96% is achieved. In addition, the robustness of the proposed method is tested across 10 different types of welding discontinuity. Moreover, the results are compared with non-optimized CNN architecture. In case of using optimized CNN the classification accuracy is improved by 10%. In addition, it is shown that based on the data which the CNN is trained with, it is possible to use this approach in real-world cases. Utilizing this method is not only improves the defect detection speed in higher volume projects but also it will improves the accuracy of defect detection in such projects.

REFERENCES

1. Rafael Vilar, Juan Zapata, and Ramón Ruiz. An automatic system of classification of weld defects in radiographic images. *Ndt & E International*, 42(5):467–476, 2009.
2. Philip Valentin. WELD CLASSIFICATION BASED ON GREY LEVEL CO-OCCURRENCE AND LOCAL BINARY PATTERNS.
3. Domingo Mery and Miguel Angel Berti. Automatic detection of welding defects using texture features. *Insight-Non-Destructive Testing and Condition Monitoring*, 45(10):676–681, 2003.
4. Jayendra Kumar, RS Anand, and SP Srivastava. Multi-class welding flaws classification using texture feature for radiographic images. In *2014 International Conference on Advances in Electrical Engineering (ICAEE)*, pages 1–4. IEEE, 2014.
5. Jayendra Kumar, RS Anand, and SP Srivastava. Flaws classification using ANN for radiographic weld images. In *2014 International Conference on Signal Processing and Integrated Networks (SPIN)*, pages 145–150. IEEE, 2014.
6. Juan Zapata, Rafael Vilar, and Ramón Ruiz. Performance evaluation of an automatic inspection system of weld defects in radiographic images based on neuro-classifiers. *Expert Systems with Applications*, 38(7):8812–8824, 2011.
7. Ioannis Valavanis and Dimitrios Kosmopoulos. Multiclass defect detection and classification in weld radiographic images using geometric and texture features. *Expert Systems with Applications*, 37(12):7606–7614, 2010.
8. Gang Wang and T Warren Liao. Automatic identification of different types of welding defects in radiographic images. *Ndt & E International*, 35(8):519–528, 2002.
9. Wenhui Hou, Ye Wei, Yi Jin, and Changan Zhu. Deep features based on a DCNN model for classifying imbalanced weld flaw types. *Measurement*, 131:482–489, 2019.
10. Domingo Mery, Vladimir Riffo, Uwe Zscherpel, German Mondragón, Iván Lillo, Irene Zuccar, Hans Lobel, and Miguel Carrasco. GDXray: The database of X-ray images for nondestructive testing. *Journal of Nondestructive Evaluation*, 34(4):42, 2015.
11. Wenming Guo, Huifan Qu, and Lihong Liang. WDXI: The Dataset of X-Ray Image for Weld Defects. In *2018 14th International Conference on Natural Computation, Fuzzy Systems and Knowledge Discovery (ICNC-FSKD)*, pages 1051–1055. IEEE, 2018.
12. M Naddaf-Sh, SeyedSaeid Hosseini, Jing Zhang, Nicholas A Brake, Hassan Zargarzadeh, et al. Real-Time Road Crack Mapping Using an Optimized Convolutional Neural Network. *Complexity*, 2019, 2019.

5 Weld Defect Analysis via Digital Radiography and Deep Learning

M-Mahdi Naddaf-Sh, Sadra Naddaf-Sh, Hassan Zargarzadeh, Sayyed M Zahiri, Maxim Dalton, Gabriel Elpers, Amir R Kashani

CHAPTER SUMMARY

Continuous and digitized monitoring and automated inspection are key parts of modern manufacturing and sustainment of aging infrastructure. The growing demand for these needs and shortage of required skill sets can slow down the global economy by increasing the risk or costs associated with catastrophic events. The diversity of requirements and specialized standards and codes around the world, along with the time-sensitive aspect of such inspections, makes automated fault detection and classification a prime application for utilizing Artificial Intelligence (AI) as an assistive tool that not only automates repeated tasks, but also provides users with supporting inference to increase confidence before and during the inspection operation. In most critical cases, a non-destructive test (NDT) must be done once immediately after the weld is created and then on a scheduled or unscheduled repeated basis as the weld ages. One of the most commonly used NDT methods is radiography imaging using penetrating gamma or X-ray radiation. Existing Assisted Defect Recognition (ADR) tools in the literature are heavily focused on high-quality X-ray images and laboratory focused imaging parameters, which in many cases are not representative of imaging done in real-world applications. Moreover, the literature has focused on welds that have undergone aging and have very clear defects, problems which classical image processing could easily address. This chapter and demonstration reviews the application of deep learning to find defects in newly created welds with minimal defect size and field-quality manually done welds as opposed to laboratory welds and address the industry standards for the classification of discontinuities (defects). First, the work developed and contextualized over 100,000 X-ray images from various welds and annotated them with a group of NDT experts with varying years of experience. Based on this data and annotations, an optimized convolutional neural network (CNN) was designed and trained for detecting discontinuity and defects. Performance of the designed CNN was tested against other CNN architectures, and the overall accuracy of 96% overall classes was achieved.

DOI: 10.1201/9781003766575-5

5.1 INTRODUCTION

One of the foremost concerns of modern-day industrial and infrastructure world is safety [74]. Whether it be in the context of towering skyscrapers, bridges or pipelines, the consequences of overlooked or undetectable mistakes can be incredibly catastrophic [38]. To ensure safety you not only have to devise and implement reliable systems, equipment and methods, but you must also have an accurate means of recognizing the presence of unexpected and potentially fatal flaws. This is because the practice of automating machinery for complex systems can be incredibly difficult and error prone [5]. Perhaps even more important, no matter how precise and consistent equipment otherwise may be, if human beings are allowed to contribute directly in the process there is always room for mistakes. As technology advances and gives rise to new ways to minimize risk for human operators and maximize efficiency or capacity, it also becomes increasingly difficult to identify the fault modes of supremely precise instrumentation [4].

Reliability of large-scale infrastructure are heavily dependant on reliability of welding massive metal structures together to function as an individual unit [23]. Naively the process is simple, apply sufficient heat to melt two pieces of metal which are fused together after they cool and harden [39]. In general, a separate piece of metal acts as a filler material which is deposited and cooled between the two pieces in question. In practice, the process requires careful adjustment of finely tuned parameters (automatic or manually enforced) that govern an energy source powerful enough to first liquefy the metal and then deposit the filler material without compromising the integrity of either the base metal or the finished product [8]. From the standpoint of a welding operator, the slightest accidental or misjudged movement can be the difference between a successful weld and a very costly mistake.

Quality of Welded joints and their inspection assessment are critical elements of industries such as marine, chemical, and aeronautical industries [77]. Digitizing and automation of monitoring the quality of product is one of the main pillars of Industry 4.0. To achieve this goal various robotics platform to increase the consistency, quality and automation of the process have been developed including robotic platforms to perform digital radiography.

It is important to note that not all construction projects involve the same fundamental structures or conditions. Of unique interest is the process of welding in pipeline construction [69]. Rather than being concerned with welding two large metal beams at varying degrees, as would be necessary in the construction of a building, welding in pipelines is principally focused with a particular type of weld, that of girth-welding. Girth welding involves welding the circumference of two cylindrical pipe joints together so as to extend the pipeline into a single contiguous unit [16].

In practice pipelines are constructed in one of two primary environments on land and underwater. In both cases, multiple teams work in tandem to ensure that the pipeline is welded, properly inspected and potentially coated with special material to protect the pipeline from corrosion [82]. When pipelines are constructed on land the pipeline is often buried under ground after each joint is welded. Pipelines that are constructed underwater or offshore are produced on large vessels from massive segmented spools and laid underwater [47]. In the latter case it becomes especially important to protect against the effects of saltwater or electrochemical corrosion with an extra layer of coating. At the end of the day, welding in both environments requires very accurate testing to ensure long life of reliable operation because the cost of repairs skyrocket by the time leaks or major damage occur [57].

Thanks to advances in robotics engineering there are actually two primary means to perform welds in the modern day. While there are still many occasions for manual welding, automatic welding is proven to be more cost efficient, more consistent, and faster in many cases, especially in pipeline construction [7]. In this case, the ability to perform automated welds inside the pipe produces a greater quality joint that is fused both internally and externally and is often necessary because it is impossible to fit a person within small diameter pipelines. Also, at the very least, it is safer to eliminate the need for a person to travel within a pipe, let alone operate welding equipment inside of it [80].

5.1.1 WELDING PROCESS

In general, manual welding is noticeably less precise and more prone to error. Hand movements are random and often unpredictable. By contrast automated welding is designed to be performed on the basis of pre-configured parameters of operation. Complex factors due to how the weld is deposited around the pipe and effects due to gravity require careful and reproducible patterns that are nearly impossible for a person to recreate every time and almost guaranteed for high quality equipment [51]. No two welding technicians will perform a weld under the same conditions in the same way and neither will a technician reproduce exactly what was done yesterday [41]. As much as an improvement as automated is over manual welding, just in the space of automated welding, there is already a surprisingly large set of potential flaws that are classifiable and well documented [54].

It is clear that a set of methodologies ought to be established for testing the quality or properties of a weld without destroying the end product. Otherwise a major purpose for testing in the first place would be defeated. In fact, there is a field dedicated precisely to this endeavor, namely the field of non-destructive testing or NDT [12]. NDT techniques are designed to probe material with some sort of stimulus and interpret the response, either manually or automatically through software [58]. The idea is that anomalies in the welding process are impossible to detect with the naked eye, especially when they are located internally away from the visible surface. Using techniques and the research from applied physics, we have a plethora of ways at our disposal to observe how a material behaves or how something is affected indirectly through the material in question [33]. Most importantly we can do so without damaging something very costly to repair later down the line.

Not only does NDT provide a means to test material properties, it is intentionally non-destructive. This means that we can actually save on the cost of more expensive repairs as we have the means to preemptively repair a given weld before it becomes too costly and ensure the quality of welds without any harm or foul should they be in flawless condition [67]. Furthermore, testing in general is necessary because it would be fundamentally impossible to easily detect flaws that are known to be problematic in practice. This is due to the fact that they are embedded deep in the material, hidden from sight [66]. Certain techniques also afford a level of precision that allows for measured judgment calls based on acceptability standards [70].

While it is clear that a certain weld may contain a defect, it is also possible through these NDT techniques such as AUT and RTR to actually measure the geometric properties of the defect [20]. In this way, one can assess whether the defect is really worth a full weld re-attempt or repair or if it is absolutely necessary [2]. Is it substantially long or deep? Is it likely to form the basis of a future crack that will eventually and completely undermine the integrity of the weld? These questions are now possible to answer because the entire material, along with the conditions and characteristics that gave rise to the defects are fully intact and explorable. Alternative methods might either weaken the material, destroy the possibility of assessing defect characteristics, or disallow the ability to test a specific weld altogether [64].

In welding, NDT ranges across a wide spectrum of techniques built on a potentially very different physical phenomenon [32]. For instance, the same fundamental ultrasound technology that enables us to observe an unborn fetus in its mother's womb can be applied fundamentally to weld inspection [37]. In NDT this is called AUT or automated ultrasonic technology. In pipeline girth (circumferential) welding, AUT can be used to identify defects based on irregularities in the patterns of waves that reflect from surfaces in the pipe [60]. AUT technology itself has many forms and variations that adjust characteristics such as the patterns of the wave pulses and the orientation or number of transducers [3].

Another perhaps even more familiar medical technique which has a similar counterpart in weld NDT is that of X-ray scanning. Here, high energy radiation is deposited directly onto material in front of a screen which shows visual contrast between regions where the material is dense and very little of the radiation is able to pass through and regions where the material is relatively less dense where the screen absorbs the radiation. This technology corresponds to a technique called RT which stands for radiographic testing [40]. In RTR, real-time radiography the same fundamental RT technology is employed but the radiation absorption pattern is captured electronically rather than on a film [46]. Many other techniques exist, such as electromagnetic or eddy current testing (ET), magnetic particle testing (MT), acoustic emission testing (AE) and the list goes on [10].

5.1.2 DIGITAL RADIOGRAPHY

Compared to other NDT techniques, RT/RTR remains a popular mainstay of inspection in practice [61]. Other techniques might be too new to have a real presence and trustworthiness among both practitioners and project stakeholders, although this is changing quickly as other techniques are gaining widespread acceptance in the inspection community [13]. Also, they might be too costly to maintain or support in the field [22]. In some cases they might even be applicable only in limited circumstances or too uninformative to form the basis of support for or against critical judgment calls [27]. Similar to most other NDT techniques, the discipline of RTR is divided into two primary responsibilities: using equipment to scan the weld and properly interpreting the scan. While there is always technological advancement in the efficient and high-quality capture of radiographic weld scans the most important task remains the latter, as this is what determines the perceived weld quality [35].

In practice RT is a very well-established technique. Entire codes or specifications for the practice of and techniques of inspection have been developed as an attempt to standardize the practice of reliable interpretation and operation [73]. One of the most well-defined standards relate to the interpretation of radiographic weld scans. In these images, experienced operators are trained to visually detect a whole host of defects which are caused by different factors related to the material and welding process. To the untrained eye, defects are difficult to classify, even imperceptible, unless told exactly what to look for. Through years of experience, individual operators learn how to identify, classify and prioritize these defects, in accordance with one of the major inspection codes which provide visual examples, descriptions and guidelines [18].

Even today, almost the entire process, with the exception of image capture itself, is managed and performed by people. People are both susceptible to visual fatigue and uniquely trained over long periods of time with an exclusive skillset that takes years to transfer to the next generation of qualified technicians [36]. A fundamental question that we can ask regarding RTR inspection is what role, if any, could the interpretation of radiographic scans be delegated to a machine? The idea of allowing a machine to assist a human operator in the process of interpreting responses from the application of NDT inspection techniques for the purpose of identifying problematic flaws is known as assisted defect recognition or ADR [83]. The key is that while we still ensure that human beings are fully and directly in control of the final assessment people are still willing and able to use whatever tools in their disposal they have to aid them in a task so difficult to consistently reproduce or perform perfectly.

Human operators, depending on their role and level of skill are expected to visually parse up to hundreds of weld scan images daily. This is a lot to ask from the eyes and the brain and is bound to at least introduce the possibility of reasonable errors that are unfortunately essentially unacceptable from a standard of quality [52]. As the process of radiographic scanning is nearly fully automated in practice, people, though specially trained, are expected to keep up with the same pace and operate with consistent and high accuracy [15]. This is at least one reason why it seems reasonable that people should be assisted in the process of interpretation and analysis of radiographic weld scans.

Given that a single individual is incapable of being at thousands of places in the world at once, we must employ the talents of many individuals to process weld images from multiple projects. Moreover, each individual has with unique project backgrounds, levels of experience, predispositions, preferences, opinions [65]. Despite the presence of NDT inspection codes and standard, it is almost totally unexpected to expect all of these individuals to agree upon the same internal standard formed over time and through experiences [78]. Even if this were not the case, it

guaranteed that individuals trained on the basis of separate codes, which are not entirely
consistent with one another, will differ with one another in the interpretation of certain key conditions [49]. It takes substantial time for someone to internalize an external standard and develop the ability to apply it as second nature professionally. If an operator is trained under the API 1104 inspection standard for pipeline welding, they do not automatically acquire the knowledge embodied in the ASME standard [50]. Even more pertinent, they must contradict themselves relative to a previous standard on occasion to adhere to a new one. People are not often capable of maintaining multiple potentially incompatible standards mentally.

The fundamental approach is a data-driven machine learning model. The advantage of this is that the experience of technicians on former projects across a wide variety of conditions are now directly available, more than any one individual could provide, even if they had sufficient time. On some level there is no real need to acquire the understanding that an experienced inspection operator possesses to develop a system that is capable learning from them [81]. There are varying ways and degrees that a machine might assist a human by focusing on tasks that machines are proven to be successful with.

Specially designed software might simply identify the presence of anomalies or more usefully classify them into distinct categories. Operators would be able to see the most likely classification candidates. Going one step further this classification could be based on specific classes that are defined in common inspection codes. As an operator is advised through ADR, they may also be quickly directed to key references or guidelines in a given code or explicit examples encountered in previous projects. When it becomes clear that different codes deviate from one another, such a tool could be trained much like a person is trained on competing standards, far more than any one individual could master in a lifetime. In essence ADR offers the opportunity to enlist the help of all professionals everywhere in every decision, even when they are completely unavailable [53].

As the inspection community gains increased faith in the accuracy and efficacy of such machine learning models and associated software tooling the possibilities for safe and valuable technological aids in the process of NDT weld inspection and analysis grow accordingly.

The focus of this research is at the intersection of most of the previously discussed key concepts. In particular, the goal was the development of a practically valuable and useful model for defect recognition in radiographic image scans on pipeline girth-welds to assist inspection technicians and operators in visual analysis. In order to accomplish this, deep-learning techniques were employed for automatically classifying defects built from a library of a large number of labeled annotations. The value of a data-driven model is rooted in the ability to learn from professional experts directly and efficiently [29]. Additionally, a deep-learning model, provided enough training data, is capable of extracting highly complex features that intuitively correspond to the same features for which professional inspection technicians develop a reliable intuition [71]. This intuition, by its very nature is unfortunately at the same time difficult to explain to another person. Even without years of professional experience, it becomes possible translate the collective experience of the inspection expert community into a language that they could understand.

Non-Destructive Testing (NDT) using radiography(RT) is one of the oldest techniques evaluate welding quality, which is used predominantly in inspection of welded joints. Utilization of RT techniques is inevitable to certify the safety and reliability of manual welds these structures. However, the process of assessment of X-ray images is both time-consuming and at times can be subjective for the expert operators due to various reasons [76]. The advent of novel image processing techniques and pattern recognition has been applied to accelerate the process of improving the image quality to help weld defect diagnosis and increase the accuracy of defect detection. Nonetheless, due to the complexity of the task, many efforts have been made to develop such an automated intelligent system to recognize defective welded joints.

5.2 LITERATURE REVIEW

Previous research on weld defect detection was mostly performed using techniques such as segmentation and texture features extraction and classification, one of which has reached the highest accuracy of 90.91% by D. Mery [55]. Artificial Neural Networks (ANNs) are used predominantly with enhancements on classifier description among researches. Kumar et al. [44] used gray-level co-occurrence (GLCM) texture features as ANN input and reached an accuracy of 86.1% for defect classification.

In another work[43], Kumar et al. added geometrical features and enhanced the accuracy to 87.34%. Juan Zapata [84] utilizes two neural classifiers, adaptive network-based fuzzy inference system (ANFIS) and ANN. In their work, ANN achieved accuracy of 78.9%, and using the 12 chosen geometrical features extracted from X-ray images as input, ANFIS reached an accuracy of 82.6%. To integrate both texture and geometrical features and to select only useful features, preventing an increase in computational complexity, Valavanis et al. [75] employed the sequential backward selection (SBS) technique to design the classifier of the ANN, which resulted in 85.4% accuracy.

Wang et al. [79] designed a weld flaw detection system, in which they extracted 12 different features from radiography images such as size, intensity, direction, location, and shape for the input of networks. They applied two well-known networks for detection of weld flaws: fuzzy k-nearest neighbor (fuzzy K-NN) and multi-layer perceptron (MLP) neural networks, which attained accuracy of 91.57% and 92.39%, respectively. However, traditional feature extraction methods used in these systems are not able to extract discriminant features.

The solution to improve pattern recognition in weld flaw detection is to take advantage of deep neural networks that can extract higher-level and more abstract features. Jin et al. [30] employed Deep Convolutional Neural Networks (DCNN) on three resampled datasets based on the GDXray [56] weld dataset and the highest achieved accuracy was 97.2% on the dataset using synthetic minority over-sampling technique (SMOTE). For the remaining datasets two techniques of random over-sampling (ROS) and random under-sampling (RUS) for modifying samples were applied and best case performances were 96.3% and 79.9% for each method, respectively.

Integration of computer vision methods using DCNNs shows exceptional promise for use in NDT applications, but require many images for the training process. Although utilizing CNNs improves the defect detection accuracy, there are known drawbacks in the conventional approaches for studying the defects. As an instance [48] studied application of deep convolution neural network for detection of casting defects. The authors reported 96% accuracy achieved through using an 8 layer CNN for detecting defects in X-ray images of casted object. They could improve the detection rate up to 8% by applying their CNN-based method over ANN-based method. Utilizing the proposed method not only make it possible to increase the quality but reduce scrap rate during casting process.

As reviewed above, deep learning-based methods for NDT improve false detection accuracy within X-ray images, whether using different CNN architectures or pre-trained models [31]. In [85], the authors deployed a CNN classification model to detect weld defects. The goal is detecting defects for aluminum alloy in robotic arc welding using an 11 layers CNN. In this method the authors reported 99.38% accuracy for a single type of welded material. For database, the authors used CCD camera instead of radiography images. In the industry using images of CCD camera for various types of materials in weld is not common, because it is not possible to detect in depth defects like lack of infusion. As another example, in [1] the imaging sensor is improved by utilizing UVV band visual sensor system but this method is not capable of penetrating through material deep enough to detect defects like internal cavity.

The authors in [34] studied generative adversarial nets (GAN) in weld defect detection. Based on the definition of defect, they categorized defects in three groups: cracks, porosity, and burn through. In their definition series of single dots considered as porosity, linear shadow that can be found vertical or parallel to the edge of welds defined as crack, and burn through defined as shadows of anomalous shape on X-ray images. The definition is not applicable in the real-world X-ray images and many defect classes will be neglected due to the limited bounded definition. Although, the proposed method could reach above 94%, the trained CNN results on other sets of X-ray weld datasets are not reported to check the generalization of the network.

On the other hand, most of reported works used the GDXray database, which is public [56], and the WDXI dataset [25], which is not available publicly. As depicted in Figure 5.1 (top image), defects in images of the GDXray database are handpicked and defects are easily visible due to their extreme nature. In addition, weld images in the GDXray database are limited to only 88 samples. Although GDXray has become a standard benchmark for testing the performance of different algorithms [30], defects in real-world X-ray images are not as obvious as those in the GDXray dataset. As it is shown in Figure 5.1 (bottom image), it is more challenging to detect defects in real-world X-ray images that are not hand-picked and do not contain oversize defects. Moreover, in previous studies, whether traditional image processing algorithms or deep learning based methods were used, differences in wall thickness of welded objects and scattering or waving in the weld root due to geometry of pipes and

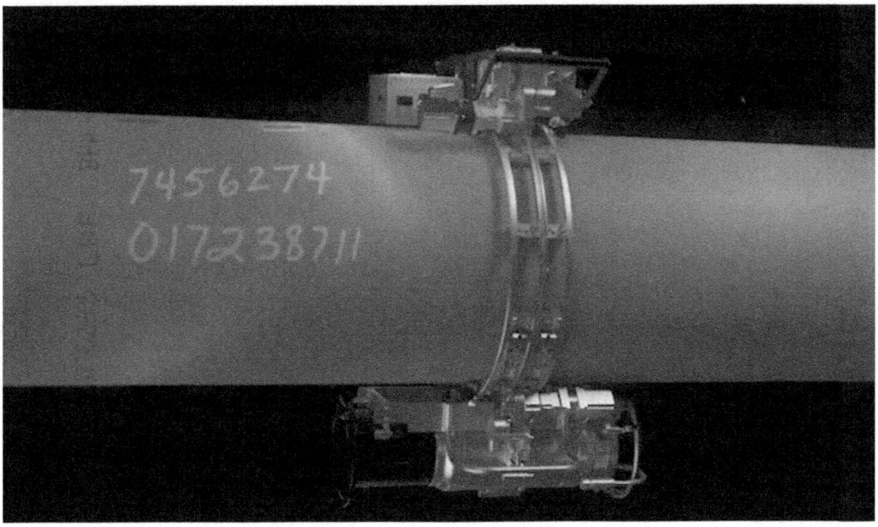

Figure 5.1 Digital X-ray detector and source on a robotic platform - Stanley Oil and Gas.

variation of welding pattern of manual welder are not present and neglected as a feature. In such cases due to change of thickness, intensity of the image varies significantly in each segment of the weld as it is shown in Figure 5.2.

In this chapter, the main goal is set to locate and identify discontinuities and defects (detection) in a realistic X-ray image dataset(SBD Dataset) and to determine the defect type. Due to the aforementioned differences with the GDXray database, previous methods either failed to detect the defects present in the SBD dataset or the overall accuracy of the methods were dramatically lowered. In order to address this problem, a deep convolutional neural network is designed and trained to be able to detect defects in SBD database.

5.3 DATABASE PREPARATION

To develop and evaluate the proposed network architecture, 5000 full-sized 15360×1024 ($Width \times Height$) pixel images of a welded pipeline containing weld defects are gathered. All full-sized images were cropped into 224×224 pixel patches, each patch including the weld's center. A total of 100,000 patches were categorized as non-defected and defected by an expert with API and ASME welding certificates. In the first phase, we selected a smaller set of non-defective and defective patches in a balanced number. This database was named SBD-1. In addition, to evaluate the network's performance against different types of defects, another database developed (SBD-2), containing more than 20,000 discontinuity patches, including 11 different types of major defects recognized by ASME and API. Defect types and quantity are

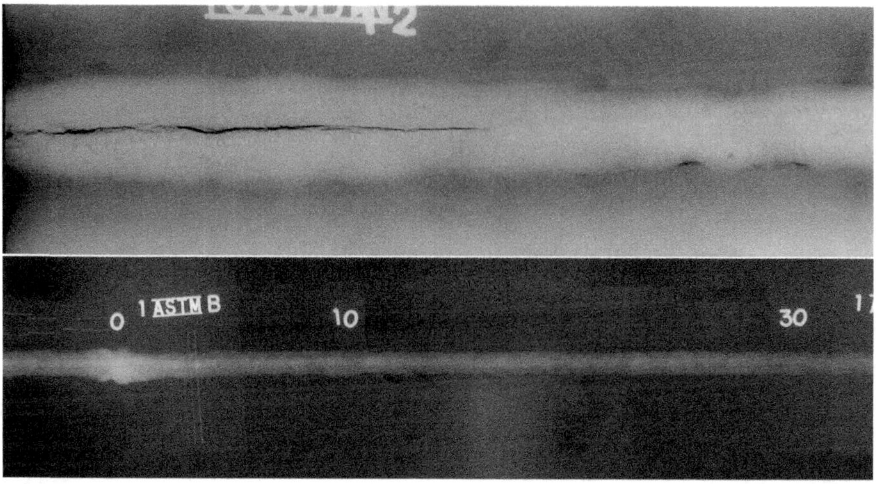

Figure 5.2 Two samples of images in GDXray database (Top) and SBD database (Bottom). The bottom image is cropped to be able to compare with the other. Defects are more visible in the GDXray database.

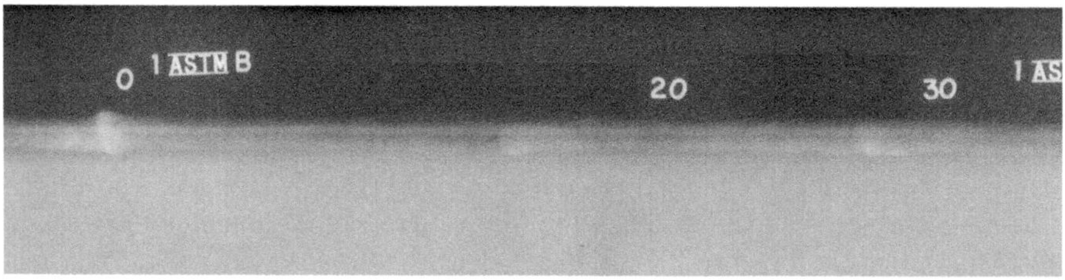

Figure 5.3 An image sample from SBD database in that two surfaces with different thicknesses are welded.

available in Table 5.1. Figure 5.4 depicts ten samples of non-defective patches, and Figure 5.5 shows 11 patches along with defect type.

5.4 EXPERIMENTAL STUDY

AlexNet[42] was first purposed in 2012 by Alex Krizhevsky et al. and was the winner of the ImageNet[14] LSVRC-2010 contest. In this work, for the first time Rectified Linear Units(RELUs) was used as activation function. The Network was trained on 1.2 million high-resolution images with the goal of classifying into 1000 categories. The 8-layer architecture of the network, which is an extension of Lenet-5[45], is consisted 5 convolutional layers followed by 3 Fully Connected ones,and Network has 60 Million parameters. As input it takes 224 RGB images, and final FC layer is a 1000-way soft-max. After that, using Deeper CNN has become prevalent among

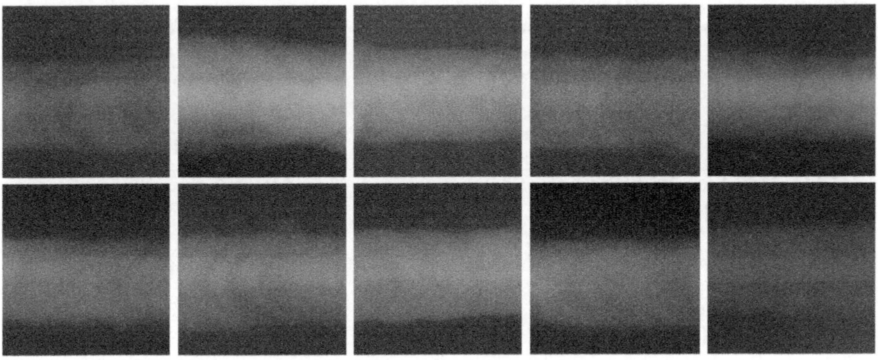

Figure 5.4 10 samples of non-defect images.

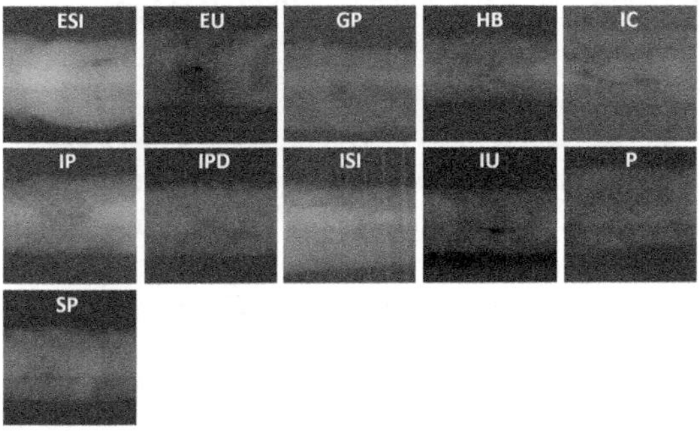

Figure 5.5 Classes of Welding Defects based on ASME standard.

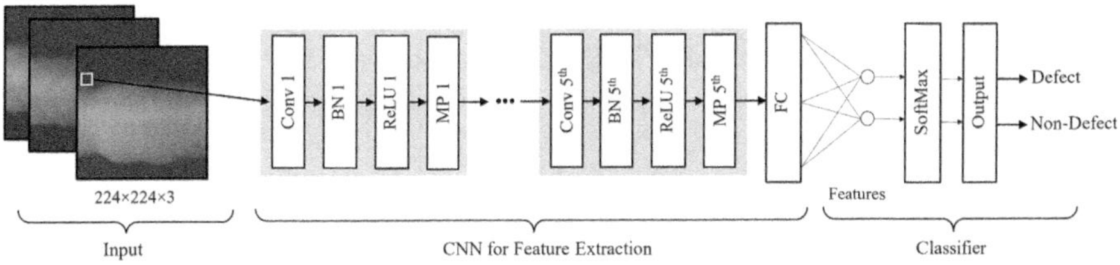

Figure 5.6 Network architecture used for weld defect detection. 5 convolution layers are used in this architecture. Conv: convolution layer, BN: batch normalization layer, MP: max-pooling layer, and FC: fully connected layer.

researchers. As a proof that deepening the network will improve model accuracy, VGG-16 and VGG-19[72] was purposed in 2014 by Karen Simonyan & Andrew Zisserman on the same ImageNet Dataset.The network has approximately 138 Million parameters. In comparison to AlexNet, they have added 11 Convolutional layer to VGG-16 and 14 to VGG-19 to improved overall accuracy of the network. In 2015 a deeper model with 152 layers, which is know as ResNet-50[28], presented by

Table 5.1
Defect Types

Defect Type	Total Number of Patches
Elongated Slag Inclusions (ESI)	5884
Hollow Bead (HB)	5204
Isolated Slag Inclusions (ISI)	3776
Gas Porosity (GP)	2528
Inadequate Penetration (IP)	1216
External Undercut (EU)	1692
Porosity (P)	785
Scattered Porosity (SP)	632
Inadequate Penetration Due to High-Low (IPD)	938
Internal Undercut (IU)	902
Internal Concavity (IC)	814

Microsoft Research. In this work, they leveraged skip connections and batch normalization to address the problem of saturation and prevent compromising of model generalization in deep networks.

Several CNN architectures like VGG-16, VGG-19, AlexNet, and ResNet were applied to the SBD-1 and SBD-2. The maximum accuracy obtained using these well-known architectures was 86% for SBD-1 and 75% for SBD-2. Because the object of interest in this problem is not a common object, i.e., weld defects, transfer learning-based methods like in [17] would not be an optimal solution. To address this problem a network architecture is proposed as depicted in 5.9. To obtain optimized values for the number of layers, filter size, and other hyperparameters (Hps), a Bayesian optimization algorithm was used [63]. In the following subsections design steps for SBD-2 database are presented. The steps will be the same for SBD-1 database with exception of number of classes.

5.4.1 DEEP LEARNING ARCHITECTURE

Figure 5.9 illustrates the overall CNN architecture including input, multiple convolutional layers, followed by a batch normalization layer, Rectified Linear Unit (ReLU) layer, max-pooling layer, a fully connected (FC), SoftMax layer and output layer for the classification task.

The first layer is the input layer that receives the input 224 × 224 patches for classification. In SBD-1 this classification would be between two classes and the SBD-2 it would be a multi-class classification problem.The defect features in each data batch were extracted using multiple convolution layers. This layer consists of various sets of neurons whose weights and biases will be updated relative to the defect features. In the convolutional layer, the neuron input consists of small sectors from the previous layer that is called the filter (kernel). The size of the filter, s_f, can be tuned from 1×1 pixels up to the size of the input image. In the convolution layer, the filter moves along the input and builds a convoluted feature map. To increase the number of feature maps, multiple filters should be used, and each filter has different weights and biases to be able to extract various features of the image. The stride (amount of horizontal and vertical movement of the filter on the input per convolution) is set to 3 pixels.

After the convolution layer, a batch normalization layer is used for reducing the CNN sensitivity to initial HPs values and decreasing training processing time. Following the batch normalization layer, a ReLU activation layer is added to apply a zero threshold to all negative values in the batch normalization layer, that means, the inputs from the previous layer b go through $max(0, b)$. The max-pooling layer downsamples the input by dividing it into rectangular pooling regions to compute the maximum of each region of gathered feature matrices. After designing the feature extractor, the FC layer is used to map the features matrix in the last layer in the form of a $1 \times c$ vector, where $c = 11$ is the number of defect types (DT) in SBD-2. For representing the probability distribution over multiple classes in the output of a classifier, a generalized model of binary logistic regression classifier (SoftMax function), is utilized after the FC [21], [9]. Considering the input of the SoftMax function as a sample defect-patch (DP) that belongs to one of the 11 DT, $DP \in DT_j$ where $j \in \{1, ..., 11\}$, then the defect patch prior probability [24] is defined as $P(DT_j)$ which shows the probability of $DP \in DT_j$ and conditional probability as $P(DP_j, \bar{\phi} | DT_j)$, where $\bar{\phi} = [\bar{\omega}, \bar{b}]$ is the parameter vector that consists of weights $\bar{\omega}$ and biases $\bar{b}$. The SoftMax function is described as:

$$S_j(DP, \bar{\phi}) = P\left(DT_j \mid DP, \bar{\phi}\right) = \frac{P\left(DP, \bar{\phi} | DT_j\right) P\left(DT_j\right)}{\sum_{m=1}^{j} P\left(DP, \bar{\phi} | DT_m\right) P(DT_m)} = \frac{\exp\left(r_j(DP, \bar{\phi})\right)}{\sum_{m=1}^{L} \exp\left(r_m(DP, \bar{\phi})\right)} \tag{5.1}$$

where $r_j(DP, \bar{\phi}) = \ln\left(P\left(DP, \bar{\phi} \mid \text{DT}_j\right) P(\text{DT}_j)\right)$ and S_j is a probability distribution as the SoftMax function output; where $0 \leq S_j \leq 1$ and $\sum_{j=1}^{13} S_j(DP, \bar{\phi}) = 1$

Following the SoftMax function, the classification output layer (cross entropy function) is used to assign each input to one of the $n = 11$ mutually exclusive defect types using the loss function shown in the following:

$$l(\bar{\phi}) = -\sum_{i=1}^{p} \sum_{j=1}^{n} d_{ij} \ln S_j\left(DP_i, \bar{\phi}\right) \tag{5.2}$$

where p is the number of samples and d_{ij} is a matrix that shows with what probability the ith sample of DP belongs to j^{th} defect type.

5.4.2 TRAINING

Stochastic Gradient Descent with Momentum (SGDM) was used to train the CNN for classification. This method updates CNN's weights and biases to minimize the loss function that measures the difference between true classified and false classified DPs.

The SGDM uses a subset of training data (mini-batch). The gradient derived from the data within the mini-batch is used for updating the weights and biases. Each update to the weights and biases is defined as one iteration. The gradient descent update law is described as:

$$\bar{\phi}_{k+1} = \bar{\phi}_k - \lambda l\left(\bar{\phi}_k\right) + \eta\left(\bar{\phi}_k - \bar{\phi}_{k-1}\right) \tag{5.3}$$

where subscript k represents the iteration number, the initial learning rate is $0 < \lambda < 1, \bar{\phi}$ is a vector that contains the weights and biases, $l(\bar{\phi})$ is the loss function and $0 \leq \eta \leq 1$ is the momentum, which defines the level of contribution from the previous step. For λ values close to the 0, the learning processed is slowed. and values close to 1 lead to either diverging or suboptimal weights. Moreover, to prevent overfitting of CNN during the training process, L2 regularization [9], [62] is utilized as follows:

$$l\left(\bar{\phi}_{k+1}\right) = l\left(\bar{\phi}_k\right) + \frac{\tau\left(\bar{w}_k^T \bar{w}_k\right)}{2} \tag{5.4}$$

where τ is the regularization factor. To address over-fitting and feature memorization, and improve the generalization of the SoftMax classifier during the training process, a modified data augmentation procedure is used during each iteration [21]; where the DPs were translated randomly in the horizontal and vertical directions by a maximum of by ± 10 pixels.

5.4.3 NETWORK HYPERPARAMETERS (HP) OPTIMIZATION

HPs in the proposed CNN architecture and SGDM are the filter size s_f, number of filters N_f, the number of CNN layers ND, η, τ, and λ. The search range for HP was defined as $ND \in \{1,2,\ldots,20\}, S_f \in \{1,2,\ldots,15\}$,$N_f \in \{1,2,\ldots,100\}, 0 \leq \eta \leq 1, 0 \leq \tau \leq 1$, and $0 < \lambda < 1$. The possible values for ND, s_f and N_f are integers, and for η, τ, and λ are logarithmically spaced values between 0 and 1. The classification error is the number of miss-classified DPs by the classifier (SoftMax).

The objective of optimization is to find optimal values for the HPs such that the classification error is minimized. So, the objective function can be considered a function with HPs as the input and the classification error as the output. Modeling of this objective function is algebraically complicated and computationally intensive. The BOA is capable of performing optimizing the HPs to minimize the classification error, while the objective function is considered as a black-box [6]. To perform the BOA, a validation set was defined that consists of 15% randomly selected DPs from the training set. The inputs of the objective function are training set and validation set.

As shown in Figure 5.7, the objective function trains the CNN and returns the classification error on the validation set. By modeling the calculated error using Gaussian process (GP) as mentioned in [19] and in multiple iterations Z, where $z = \{1,2,\ldots,83\}$, the BOA finds the optimal values for HPs that minimize the classification error. The kernel function that was used for GP is the Automatic Relevance Determination (ARD) Matéra 5/2 in [68]. In addition, the acquisition function

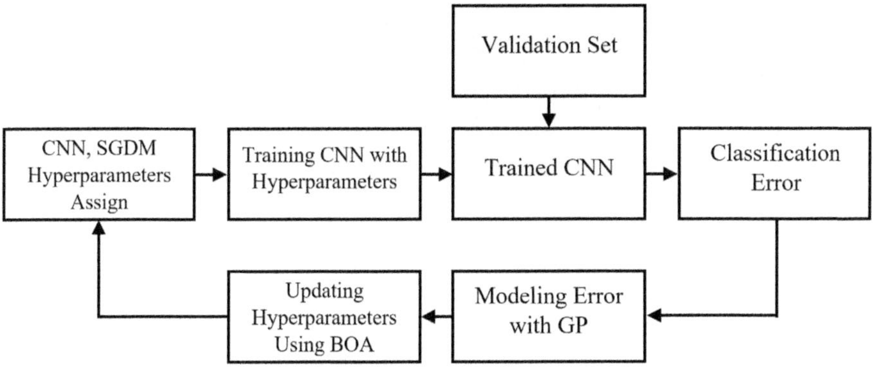

Figure 5.7 Block diagram of optimizing hyperparameters using BOA.

$(q_z(HP))$ that is used for the GP is the Expected Improvement function E(.) [59], as follows:

$$q_z(HP) = \arg\max_{HP} \mathrm{E}\left(\max\left\{0, f_{z+1}(HP) - f_z^{\max}(HP)\right\}\right) \tag{5.5}$$

where $f_z^{\max}(HP)$ is the current maximum observed value for the objective function. The next estimation for maximizing the objective function is obtained by using the acquisition function. The GP posterior is updated in each iteration using Equation 5:

$$P(f_z \mid u) = \frac{P(u \mid f_z)P(f_z)}{P(u)} \tag{5.6}$$

where $u = \{(HP_z, f_z), z = 1 : 100\}$.

The extrema of $f_z(HP)$ was obtained numerically at sampled values of the function. A closed form expression of the objective function is not required within the BOA mathematical structure [11]. The objective function and acquisition function for two of SGDM HPs, i.e., η and λ during the optimization process are shown in Figure 5.8. As depicted in Figure 5.8(a), the observed points are demarcated by blue dots $(f_z(HP))$, the model mean that is obtained from the observations is depicted as the red surface, and subsequent evaluation point addition is demarcated with a black dot. Moreover, Figure 5.8(b) illustrates the acquisition function. The objective function is shown to reach a minimum at the 53th iteration; this point is demarcated with a black star. Figure 5.8(b) shows the maximum feasible value that is generated upon minimizing the classification error.

The total number of iterations was set to 80. Each iteration calculates the classification error among 700 randomly selected DPs from 11 defect batches. The BOA was evaluated statistically using the Wald method [26] by representing the images in the test set as independent events with a known probability of success. The number of misclassified images were represented with a binomial distribution. By applying the trained CNN with optimized HPs on the test set and computing the number of

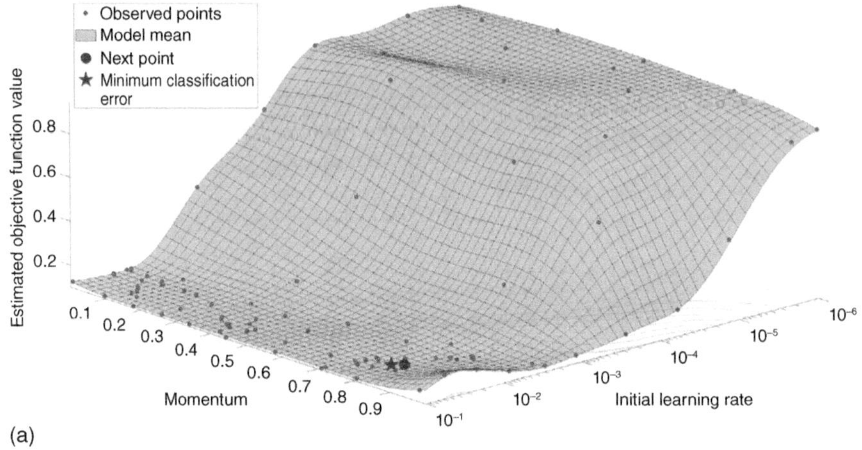

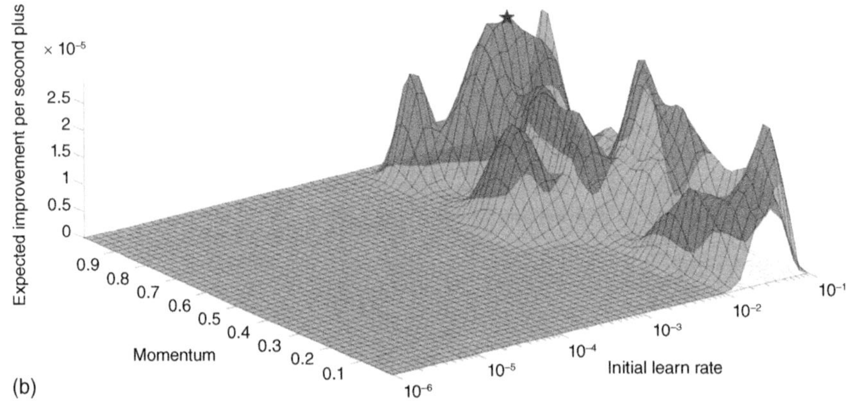

Figure 5.8 (a) The observation function model. (b) Acquisition function for two parameters of the SGDM. The starred point is the calculated optimal values at the 53th iteration for η and λ.

correctly classified DPs, the test error E_r is defined as follows:

$$E_t = 1 - \frac{1}{b}\sum_{i=1}^{b} Ds_i \tag{5.7}$$

where Ds and b are the number of correctly classified DPs, and the total number of DPs in the test set, respectively. Note to evaluate the trained CNN performance on the test set without exposing the CNN to the optimization process, E_s is used to obtain the standard error. This approach helps to increase the optimization speed.

The standard error is represented as follows:

$$E_s = \sqrt{\frac{E_t}{Ds}(1 - E_t)} \tag{5.8}$$

Moreover, as the target of this research, to obtain a ±8 error margin, a confidence interval of 92% is defined to calculate the generalization error E_G defined as:

$$E_G = E_t \pm 0.92E_s \tag{5.9}$$

The final HPs value for the CNN were $S_{f1} = 7$,$S_{f2} = 7$,$S_{f3} = 9$, $N_{f1} = 235$, $N_{f2} = 105$, and $N_{f3} = 98$. Also, the optimized values for SGDM were $\lambda = 0.0016737$,$\eta = 0.040151$ and $\tau = 0.003134$. Applying the optimal HP values to the CNN and SGDM yields a CNN with 20 layers and 91.20% accuracy in 8 epochs. Moreover, the minimized value for the loss function was 0.021 in 8 epochs. Also, the generalized error E_G interval for the test set was [0.5170.0115].

The same steps followed for the SBD-1 and the resulted HPs value for the CNN were $S_{f1} = 3$,$S_{f2} = 5$,$S_{f3} = 9$, $N_{f1} = 107$, $N_{f2} = 87$, and $N_{f3} = 159$. The optimized values for SGDM were $\lambda = 0.002128$,$\eta = 0.025287$ and $\tau = 0.003876$. Applying the optimal HP values to the CNN and SGDM yields a CNN with 20 layers and 95.86% accuracy in 8 epochs. Moreover, the minimized value for the loss function was 0.037 in 8 epochs.

5.5 EXPERIMENTAL IMPLEMENTATION

There is an increasingly valuable resource for managing both training and inference pipelines based on sophisticated models, namely large-scale cloud computing platforms. Through services that are specifically designed to offload high capacity data storage and intensive computations to scale-able remote specially-designed servers, it becomes possible to develop and serve complex models such as the aforementioned and chosen deep-learning CNN. In the course of the model's development multiple AWS services were employed to handle all stages of production, including means to easily supply the model through the input of dedicated experts. For such advanced models it simply is not possible anymore to train efficiently on local generic machines with the amount of data necessary so that the model can learn or extract sophisticated features and maximize accuracy in its predictions.

In particular, the storage service known as S3 allowed for systematic massive-scale large file data hosting. In practice, the electronically captured radiographic DICOM (Digital Imaging and Communications in Medicine) images assessed and analyzed by inspection operators are quite unwieldy to manage on local machines and essentially impossible to collect at one time. As technicians capture data through equipment designed for real time radiography these, images are directed to a database of of files ready for processing to either be used as training or testing data for the model or for the opportunity for additional annotations and detailed labelling. As these images are collected from projects that are confidential and display information that should not be seen by third parties, a mandatory redaction step is applied

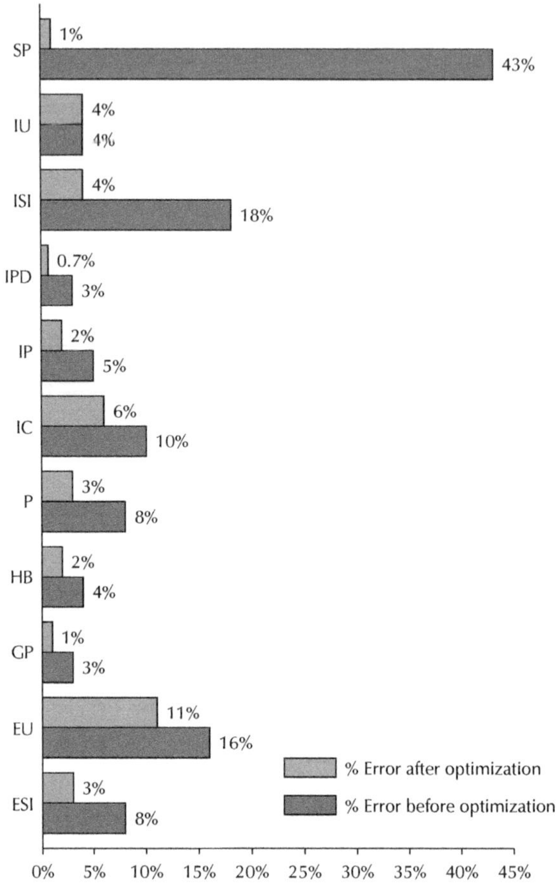

Figure 5.9 Comparison between error among 11 defect categories before and after optimization.

using optical character recognition models to maximize the accuracy of redaction for exclusively confidential information. Once these images have been redacted they are ready for labeling and annotation or further processing.

Through the power of miscroservices architecture and services that enable scalable computation effects such as AWS Lambda each processing step in the development of the model can be produced independently. A separate series of proprietary software tools were responsible for extracting images from large data storage and preparing the input in a preprocessing step for direct application as training data into the model. As these images are essentially long rectangular unwrapped scans of the weld and pipe circumference, it was necessary to employ a method for essentially extracting the region of the image where the actual weld exists between the walls of the pipe and producing specifically sized slices that follow the weld precisely.

These slices are then related to labeled defect annotations in one of two ways. First, experts in RTR inspection are asked to annotate and classify suspected defects from the slices directly or from the original image, even outside the context of the project where the weld was originally captured. Second, specific metadata in the image recording the annotations produced from proprietary inspection and annotation tooling software in the project itself is parsed and cross referenced with the exact location of the slice relative to the original image. Finally after the appropriate labels are applied to the corresponding image slices, each centered on the weld and each containing the visual pattern of the defect class given by the label, the labeled training data is applied to the model.

5.6 CONCLUSION

In this work, two realistic welding quality datasets for training deep learning models were created based on radiography images collected from various projects and NDT expert-annotated datasets SBD-1 and SBD-2. An optimized CNN was designed in order to find defects in the weldment and heat-affected zones and was subsequently trained and evaluated based on prepared datasets. An accuracy of 96% was achieved. In addition, the robustness of the proposed method was tested across 11 different types of welding discontinuity. Moreover, the results were compared with the non-optimized CNN architecture. In the case of using an optimized CNN, the classification accuracy was improved by 10%. In addition, this approach was found to be applicable to real-world datasets. This method improved not only the defect detection speed, but also the accuracy of defect detection in high-volume projects.

REFERENCES

1. Zhang, Y.; You, D.; Gao, X.; Zhang, N.; and Gao, P.P. 2019. Welding defects detection based on deep learning with multiple optical sensors during disk laser welding of thick plates. *Journal of Manufacturing Systems* 51:87–94.
2. Aljaroudi, A.; Khan, F.; Akinturk, A.; Haddara, M.; and Thodi, P. 2015. Risk assessment of offshore crude oil pipeline failure. *Journal of Loss Prevention in the Process Industries* 37:101–109.
3. Alobaidi, W.M.; Alkuam, E.A.; Al-Rizzo, H.M.; Sandgren, E.; et al. 2015. Applications of ultrasonic techniques in oil and gas pipeline industries: A review. *American Journal of Operations Research* 5(04):274.
4. Arntz, M.; Gregory, T.; and Zierahn, U. 2017. Revisiting the risk of automation. *Economics Letters* 159:157–160.
5. Backström, T., and Döös, M. 1995. A comparative study of occupational accidents in industries with advanced manufacturing technology. *International Journal of Human Factors in Manufacturing* 5(3):267–282.
6. Baptista, R., and Poloczek, M. 2018. Bayesian optimization of combinatorial structures. arXiv preprint arXiv:1806.08838.
7. Beeson, R. 1999. Pipeline welding goes mechanized. *Welding Journal (Miami)* 78(11).

8. Benyounis, K.Y.; Olabi, A.G.; and Hashmi, M.S.J. 2005. Effect of laser welding parameters on the heat input and weld-bead profile. *Journal of Materials Processing Technology* 164:978–985.
9. Bishop, C.M. 2006. *Pattern Recognition and Machine Learning*. Springer.
10. Bray, D.E., and McBride, D. 1992. Nondestructive testing techniques. *STIA* 93:17573.
11. Brochu, E.; Cora, V.M.; and De Freitas, N. 2010. A tutorial on Bayesian optimization of expensive cost functions, with application to active user modeling and hierarchical reinforcement learning. arXiv preprint arXiv:1012.2599.
12. Cartz, L. 1995. Nondestructive testing. ASM International, Materials Park, OH (United States).
13. De Raad, J.A., and Dijkstra, F.H. 1997. Mechanized ultrasonic testing on girth welds during pipeline construction. *Materials Evaluation* 55(8).
14. Deng, J.; Dong, W.; Socher, R.; Li, L.-J.; Li, K.; and Fei-Fei, L. 2009. Imagenet: A large-scale hierarchical image database. In *2009 IEEE Conference on Computer Vision and Pattern Recognition*, 248–255.
15. Domke, M.C.; Messinger, J.H.; Soorianarayanan, S.; Lambdin, T.E.; and Sbihli, S.L. 2015. Systems and methods for analyzing data in a non-destructive testing system. US Patent 9,217,999.
16. Fairchild, D.P.; Macia, M.L.; Bangaru, N.V.; Koo, J.Y.; et al. 2004. Girth welding development for X120 linepipe. *International Journal of Offshore and Polar Engineering* 14(01).
17. Ferguson, M.K.; Ronay, A.K.; Lee, Y.-T.T.; and Law, K.H. 2018. Detection and segmentation of manufacturing defects with convolutional neural networks and transfer learning. *Smart and Sustainable Manufacturing Systems* 2.
18. Forsyth, D.S.; Yolken, H.T.; and Matzkanin, G.A. 2006. A brief introduction to nondestructive testing. *AMMTIAC Q* 1.
19. Gelbart, M.A.; Snoek, J.; and Adams, R.P. 2014. Bayesian optimization with unknown constraints. arXiv preprint arXiv:1403.5607.
20. Gericke, O.R. 1963. Determination of the geometry of hidden defects by ultrasonic pulse analysis testing. *The Journal of the Acoustical Society of America* 35(3):364–368.
21. Goodfellow, I.; Bengio, Y.; and Courville, A. 2016. *Deep Learning*. MIT Press.
22. Gordon, R.; Holdren, R.; Johnson, M.; and Lozev, M. 2003. Reducing pipeline construction costs: New technologies. *Welding in the World* 47(5-6):7–14.
23. Gourd, L.M. 1986. *Principles of welding technology*. Edward Arnold London.
24. Graves, A.; Fernández, S.; Gomez, F.; and Schmidhuber, J. 2006. Connectionist temporal classification: labelling unsegmented sequence data with recurrent neural networks. In *Proceedings of the 23rd International Conference on Machine Learning*, 369–376.
25. Guo, W.; Qu, H.; and Liang, L. 2018. WDXI: The Dataset of X-Ray Image for Weld Defects. In *2018 14th International Conference on Natural Computation, Fuzzy Systems and Knowledge Discovery (ICNC-FSKD)*, 1051–1055.
26. Harrell Jr, F.E. 2015. *Regression Modeling Strategies: with Applications to Linear Models, Logistic and Ordinal Regression, and Survival Analysis*. Springer.
27. He, Y.; Pan, M.; Luo, F.; and Tian, G. 2011. Pulsed eddy current imaging and frequency spectrum analysis for hidden defect nondestructive testing and evaluation. *NDT & E International* 44(4):344–352.
28. He, K.; Zhang, X.; Ren, S.; and Sun, J. 2016. Deep residual learning for image recognition. In *Proceedings of the IEEE Conference on Computer Vision and Pattern Recognition*, 770–778.

29. Henn, M.-A.; Zhou, H.; and Barnes, B.M. 2019. Data-driven approaches to optical patterned defect detection. *OSA Continuum* 2(9):2683–2693.
30. Hou, W.; Wei, Y.; Jin, Y.; and Zhu, C. 2019. Deep features based on a DCNN model for classifying imbalanced weld flaw types. *Measurement* 131:482–489.
31. Hou, W.; Zhang, D.; Wei, Y.; Guo, J.; and Zhang, X. 2020. Review on Computer Aided Weld Defect Detection from Radiography Images. *Applied Sciences* 10(5):1878.
32. Howard, D.C. 1978. Non-destructive testing of pipeline. US Patent 4,098,126.
33. Hunaidi, O.; Bracken, M.; and Wang, A. 2008. Non-destructive testing of pipes. US Patent 7,328,618.
34. Zhang, H.; Chen, Z.; Zhang, C.; Xi, J.; and Le, X. 2019. Welddefect detection based on deep learning method. In *2019 IEEE 15th International Conference on Automation Science and Engineering (CASE)*, 1574–1579.
35. Jarmulak, J.; Kerckhoffs, E.J.H.; and van't Veen, P.-P. 2001. Case-based reasoning for interpretation of data from non-destructive testing. *Engineering Applications of Artificial Intelligence* 14(4):401–417.
36. Jarvis, J.F. 1979. Visual inspection automation. In *COMPSAC 79. Proceedings. Computer Software and The IEEE Computer Society's Third International Applications Conference, 1979*, 251–255.
37. Jensen, J.A. 2007. Medical ultrasound imaging. *Progress in Biophysics and Molecular Biology* 93(1-3):153–165.
38. Jeong, B.Y. 1998. Occupational deaths and injuries in the construction industry. *Applied Ergonomics* 29(5):355–360.
39. Khan, M.I. 2007. *Welding science and technology*. New Age International.
40. Knight, S., and Drake, S.G. 2014. X-ray inspection apparatus for pipeline girth weld inspection. US Patent 8,923,478.
41. Kobayashi, K.; Ishigame, S.; and Kato, H. 2003. Skill training system of manual arc welding. In *Entertainment Computing*, 389–396. Springer.
42. Krizhevsky, A.; Sutskever, I.; and Hinton, G.E. 2012. Imagenet classification with deep convolutional neural networks. In *Advances in Neural Information Processing Systems*, 1097–1105.
43. Kumar, J.; Anand, R.S.; and Srivastava, S.P. 2014. Flaws classification using ANN for radiographic weld images. In *2014 International Conference on Signal Processing and Integrated Networks (SPIN)*, 145–150.
44. Kumar, J.; Anand, R.S.; and Srivastava, S.P. 2014. Multi-class welding flaws classification using texture feature for radiographic images. In *2014 International Conference on Advances in Electrical Engineering (ICAEE)*, 1–4.
45. Lecun, Y.; Bottou, L.; Bengio, Y.; and Haffner, P. 1998. Gradient-based learning applied to document recognition. *Proceedings of the IEEE* 86(11):2278–2324.
46. Light, G.M. 1988. Demonstration of realtime radiography on pipeline girth welds. Technical report, Southwest Research Inst., San Antonio, TX (USA). Nondestructive Evaluation.
47. Lin, T.R.; Guo, B.; Song, S.; Ghalambor, A.; and Chacko, J. 2005. *Offshore pipelines*. Elsevier.
48. Lin, J.; Yao, Y.; Ma, L.; and Wang, Y. 2018. Detection of a casting defect tracked by deep convolution neural network. *The International Journal of Advanced Manufacturing Technology* 97(1–4):573–581.
49. Lumb, R.F. 1969. Non-destructive testing of high-pressure gas pipelines. *Non-Destructive Testing* 2(4):259–268.

50. Ma, B.; Shuai, J.; Wang, J.; and Han, K. 2011. Analysis on the latest assessment criteria of ASME B31G-2009 for the remaining strength of corroded pipelines. *Journal of Failure Analysis and Prevention* 11(6):666–671.
51. Matsutani, T.; Miyasaka, F.; Oji, T.; and Hirata, Y. 1997. Mathematical modelling of GTA girth welding of pipes. *Welding International* 11(8):615–620.
52. Megaw, E.D. 1979. Factors affecting visual inspection accuracy. *Applied Ergonomics* 10(1):27–32.
53. Meier, J.; Tsalicoglou, I.; and Mennicke, R. 2017. The future of NDT with wireless sensors, AI and IoT. In *APCNDT Conf. Proc. ID273/3962.*
54. Mendez, P.F., and Eagar, T.W. 2003. Penetration and defect formation in high-current arc welding. *Welding Journal* 82(10):296.
55. Mery, D., and Berti, M.A. 2003. Automatic detection of welding defects using texture features. *Insight-Non-Destructive Testing and Condition Monitoring* 45(10):676–681.
56. Mery, D.; Riffo, V.; Zscherpel, U.; Mondragón, G.; Lillo, I.; Zuccar, I.; Lobel, H.; and Carrasco, M. 2015. GDXray: The database of X-ray images for nondestructive testing. *Journal of Nondestructive Evaluation* 34(4):42.
57. Mgonja, C. 2017. The consequences of cracks formed on the oil and gas pipelines weld joints. *Int. J. Eng. Trends Technol.* 54:223–232.
58. Mix, P.E. 2005. *Introduction to nondestructive testing: a training guide*. John Wiley & Sons.
59. Mockus, J. 2012. *Bayesian Approach to Global Optimization: Theory and Applications*, volume 37. Springer Science & Business Media.
60. Morgan, L.; Nolan, P.; Kirkham, A.; and Wilkinson, R. 2003. The use of automated ultrasonic testing (AUT) in pipeline construction. *Insight-Non-Destructive Testing and Condition Monitoring* 45(11):746–753.
61. Morgan, L. 2018. Testing defects in automated ultrasonic testing and radiographic testing. *Insight-Non-Destructive Testing and Condition Monitoring* 60(11):606–612.
62. Murphy, K. 2013. Machine learning: a probabilistic approach. *MIT Press* 1163:1169–1175.
63. Naddaf-Sh, M.; Hosseini, S.; Zhang, J.; Brake, N.A.; Zargarzadeh, H.; et al. 2019. Real-Time Road Crack Mapping Using an Optimized Convolutional Neural Network. *Complexity* 2019.
64. Nestleroth, J.B. 2006. Pipeline in-line inspection challenges to ndt. *Insight-Non-Destructive Testing and Condition Monitoring* 48(9):524–524.
65. Norros, L. 1998. Human and organisational factors in the reliability of non-destructive testing (NDT). *The Finnish Research Programme on the Structural Integrity of Nuclear Power Plants*, 271.
66. Pisarski, H.G., and Wignall, C.M. 2002. Fracture toughness estimation for pipeline girth welds. In *International Pipeline Conference*, volume 36207, 1607–1611.
67. Quirk, J. 1999. Achieving greater efficiency in NDT inspections. *Sensor Review*.
68. Rasmussen, C.E. 2003. Gaussian processes in machine learning. In *Summer School on Machine Learning*, 63–71. Springer.
69. Roe, W.G. 1966. Welding system. US Patent 3,278,721.
70. Seto, A.; Masuda, T.; Machida, S.; and Miki, C. 2000. Very low cycle fatigue properties of butt welded joints containing weld defects. Study of acceptable size of defects in girth welds of gas pipelines. *Welding International* 14(1):26–34.

71. Shaheen, F.; Verma, B.; and Asafuddoula, M. 2016. Impact of automatic feature extraction in deep learning architecture. In *2016 International Conference on Digital Image Computing: Techniques and Applications (DICTA)*, 1–8.
72. Simonyan, K., and Zisserman, A. 2014. Very deep convolutional networks for large-scale image recognition. arXiv preprint arXiv:1409.1556.
73. Snyder, S.T. Alternative Acceptance Criteria for Pipeline Girth Welds.
74. Törner, M., and Pousette, A. 2009. Safety in construction–a comprehensive description of the characteristics of high safety standards in construction work, from the combined perspective of supervisors and experienced workers. *Journal of Safety Research* 40(6):399–409.
75. Valavanis, I., and Kosmopoulos, D. 2010. Multiclass defect detection and classification in weld radiographic images using geometric and texture features. *Expert Systems with Applications* 37(12):7606–7614.
76. Valentin, P. WELD CLASSIFICATION BASED ON GREY LEVEL CO-OCCURRENCE AND LOCAL BINARY PATTERNS.
77. Vilar, R.; Zapata, J.; and Ruiz, R. 2009. An automatic system of classification of weld defects in radiographic images. *NDT & E International* 42(5):467–476.
78. Wall, M. 2017. Human factors guidance to improve reliability of non-destructive testing in the offshore oil and gas industry. In *7th European-American Workshop on Reliability of NDE.*
79. Wang, G., and Liao, T.W. 2002. Automatic identification of different types of welding defects in radiographic images. *NDT & E International* 35(8):519–528.
80. Wang, R., and Guo, R.-J. 2011. Developments of automatic girth welding technology in pipelines. *Dianhanji/Electric Welding Machine* 41(9):53–55.
81. Williams Jr, D.R.; Gutzwiller, L.R.; Hazen, M.U.; Anderson, B.S.; McIntyre, A.; and Abeles, T. 2016. Classifying data with deep learning neural records incrementally refined through expert input. US Patent 9,324,022.
82. Yapp, D., and Blackman, S.A. 2004. Recent developments in high productivity pipeline welding. *Journal of the Brazilian Society of Mechanical Sciences and Engineering* 26(1):89–97.
83. Yella, S.; Dougherty, M.S.; and Gupta, N.K. 2006. Artificial intelligence techniques for the automatic interpretation of data from non-destructive testing. *Insight-Non-Destructive Testing and Condition Monitoring* 48(1):10–20.
84. Zapata, J.; Vilar, R.; and Ruiz, R. 2011. Performance evaluation of an automatic inspection system of weld defects in radiographic images based on neuro-classifiers. *Expert Systems with Applications* 38(7):8812–8824.
85. Zhang, Z.; Wen, G.; and Chen, S. 2019. Weld image deep learning-based on-line defects detection using convolutional neural networks for Al alloy in robotic arc welding. *Journal of Manufacturing Processes* 45:208–216.

6 Explainable Real-Time Object Detection in 2D Radiography

Sadra Naddaf-Sh, M-Mahdi Naddaf-Sh, Hassan Zargarzadeh, Maxim Dalton, Soodabeh Ramezani, Gabriel Elpers, Vinay S Baburao, Amir R Kashani

CHAPTER SUMMARY

Quality Inspection and defect detection play a critical role in infrastructure safety and integrity specially when it comes to aging infrastructure mostly owned by governments around the world. One of the prevalent inspections performed in the industry is non-destructive testing (NDT) using radiography imaging. Growing demand, shortage of experts, diversity of required skills, and specific regional standards with a time-limited requirement of inspection results make automated inspection an urgent need. Therefore, utilizing Artificial Intelligence (AI)-based tools as an assistive technology, has become a trend for industrial applications, which automates repeated tasks and provides increased confidence before and during the inspection operation. Most of the works in quality assessment are focused on the classification of few categories of defects and mostly performed on public or non-comprehensive research datasets. In this work, a scalable, efficient, and real-time deep learning family of models for detection and classification of 10 various categories of weld characteristics on a real-world industrial dataset is presented. The Models are evaluated and compared against each other, various critical hyperparameters and components are optimized, and local explainability of models are discussed. Additionally, AutoAugment for object detection and various techniques are utilized and investigated. The best performance for object detection and classification for 10 class models is reached by mean average precision of 72.4% and top-1 accuracy of 90.2%, respectively. Also, the fastest object detection model is able to evaluate a full 15360×1024 pixels weld image in 0.39 seconds. Finally, the proposed models are deployable on edge-devices to perform as assistant to NDT experts or auditing professionals.

6.1 INTRODUCTION

Inspection and assessment of welded joints are critical in many industries such as marine, aerospace, chemical, and specifically in oil and gas industries [1]. Welded joints are among vulnerable parts of any industrial infrastructure including pipelines. Hence, preliminary weld inspection during the construction has a crucial role in its

DOI: 10.1201/9781003766575-6

longevity as a small discontinuity can grow into an utter failure over time [2, 3]. Moreover, pipeline failures can damage life in large-scale and is a threat to the environment [4]. Furthermore, it is very costly to maintain continuous inspection to track the growth of initial imperfections over time or efforts to restore the surrounding environment or the pipeline when defects are larger than a certain threshold [5, 6]. Thus, weld inspection is the most economical preventive approach specifically at early stages of its construction. Among different Non-Destructive Testing (NDT) technologies at the point of constructions, Radiographic Testing (RT), Ultrasonic Testing (UT), and Magnetic Testing (MT) are of great importance. Currently RT, in which X-ray imaging of the welded part is done, is preferred due to the universal training and accuracy of its technology [7]. Nonetheless, analysis of X-ray images is time-consuming, tedious, and at the end different experts might have different opinions and hence auditing is essential [8]. Thus, automation of these systems is of interest in the industry to certify reliability and safety of the product in various stages of construction, approval, audit and risk assessment.

In recent decades, many research have been conducted on automation of tasks employing robots [10, 11, 12, 13], including robotic platforms automation of welding operation to accelerate the process and reduce human error. As an instance, Fig. 6.1 shows a robotic Digital X-ray photographer by Stanley Oil and Gas. The robot autonomously conducts the X-ray imaging that significantly minimizes the human intervention to prevent the operators from exposure [9]. After a robotic imaging process is done, human experts use images generated to inspect the welds. However, recent rapid improvement in Machine Learning, Computer Vision, and Pattern Recognition has opened new roads to provide novel solutions in order to address the challenges regarding ultimate defect diagnosis and complete tractability of discontinuities over the pipeline's life-cycle [2, 3, 8, 14, 15]. In the following, a review on related research performed in weld and defect diagnosis is provided.

Previous research works with focus on defect analysis are mainly divided into two smaller subgroups. Before the prevalence of Deep Learning and Convolutional Neural Networks (CNNs) approaches in the early 2010s, procedures focused on traditional image processing methods for image pre-processing and classification utilizing classical machine learning methods (e.g., Support Vector Machine (SVM)) and training Artificial Neural Networks (ANNs) based on hand-crafted features extracted from image patches (cropped rectangular pieces of a larger image). Among these works, they mainly focused on classifying defect and non-defect images and assigning a single label to an image patch with or without segmentation of defect area. D. Mery and M. Berti [16] used texture features to train ANNs and the best result reached 8% false alarm. In [17], Gray Level Co-occurrence Matrix (GLCM) texture features used for multi-class ANNs with 86.1% accuracy, and optimized to reach 87.3% by applying Levenberg–Marquardt optimizing function in [18]. A similar approach to classify defects with a combination of statistical and geometric features and utilizing Top-Hat filtering, Thresholding, Morphological smoothing as pre-processing is presented in [19] resulted in 91% accuracy in detecting defects and non-defects and 96% in classifying of a hundred of test images containing low

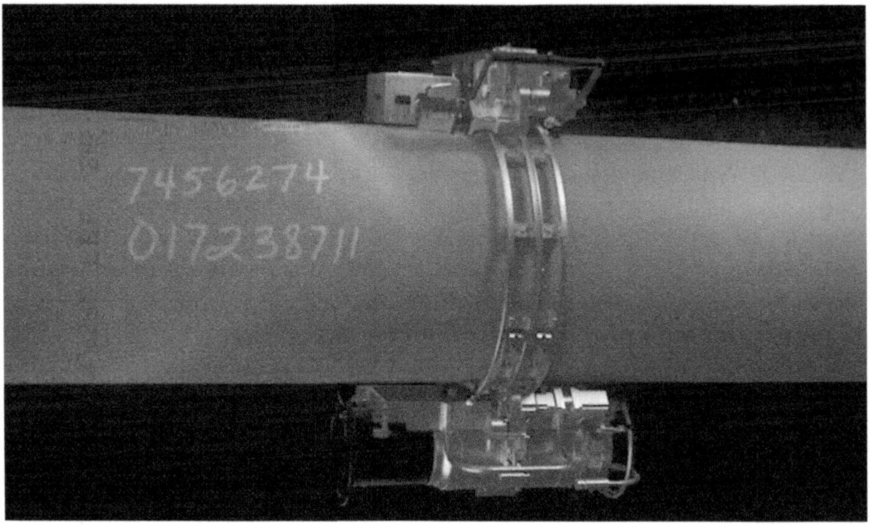

Figure 6.1 Digital X-ray detector and source on a robotic platform - Stanley Oil and Gas.

contrast images. In [20], Wiener filter is considered the best enhancement as it leads to lower Rooted Mean Square Error (RMSE) in comparison with Median filtering and Contrast enhancement, and also defective segments are obtained from the segmented image using an automatic threshold. Finally, for feature extraction, the lexicoronically one-dimensional signal of the image is generated, Mel-Frequency Cepstral Coefficients (MFCCs) and polynomial coefficients are extracted from the Power Density Spectra (PDSs) of the image and passed into ANN, which reduced false positive rate to 7%. Lim al.[21] employed a Multi-Layer Perceptron (MLP) network trained on a simulated dataset of weld radiographic images for classification of the patches.

Juan Zapata in [22] used an Adaptive Network-based Fuzzy Inference System (ANFIS) and ANN, in which geometrical and texture features were selected with respect to minimizing computational complexity and reached 82.6% accuracy. I. Valavanis and D. Kosmopoulos [23] applied certain classifiers for distinguishing between six types of defects annotated based on British Standards or labeling as non-defect. Pre-processing steps of their research include utilizing local threshold, graph-based segmentation, and then geometric and texture features are used as input for classifiers like ANN, K-Nearest Neighbor (KNN) and SVM. In [24], a comprehensive review of similar methods is provided. It can be concluded that classical approaches require major pre-processing steps before feature extraction and pre-processing enhancements have direct impact on final accuracy.

On the other hand, a few research focused on image segmentation to provide a general understanding of defect localization. M. Carrasco and D. Mery [25], presented a method for segmenting defects. The method consists of a few steps: median filtering, Bottom-hat filter, binary thresholding, and Watershed transform. The results suggested an Area Under Curve (AUC) of 93.58% for ten images. In [26], sliding window approach is used for weld object detection based on a large set of features. In [27], Gharsallah et al. proposed a method to address non-robustness of defect segmentation caused by uneven illumination, based on level set active contour guided with an off-center saliency map, in which an energy function gets minimized to achieve segmentation. Despite faster convergence and higher accuracy than Local Image Filtering and Contrast enhancement, the method requires further investigation to minimize human intervention in finding Region of Interest (ROI). In [28], defect segmentation problem is addressed using Gabor filtering and Canny edge detector. As a more recent research, which is also evaluated on aerospace weld dataset, a novel pixel-wise segmentation defect detection system is presented in [8]. Dong et al. [8] described a system to detect weld defects by using Random Forest instead of Softmax as the classifier of a U-net [29]. The approach is pixel-wise labeling of highly similar circular defects, which are prevalent in aerospace industries.

Since the prevalence of Deep Convolutional Neural Networks (DCNN), many works have focused on using these models for feature extraction/selection instead of traditional hand-crafted feature extraction and non-robust methods. Primarily two general tasks are performed using DCNNs (i.e., classification and object detection task). Furthermore, weld defect dataset has class-imbalance issue, since the number of weld defects might not distribute equally among different classes. Hoe et al. [30] focused on extending three types of datasets using Auto Encoders to address the imbalance problem. Next, a few models, including DCNNs and other models based on extracted features are trained to classify four different types of defects and reached accuracy of 97.2%. Ajmi et al. [32] explored two-class (Porosity and Lack of Penetration) classification of weld defects. Data augmentation through horizontal mirroring, translations, and RGB channels modification are applied to boost model performance, and 85.2% accuracy is reported with transfer learning utilizing AlexNet [33] and addition of a few drop-out layers as well as modified final layer on GDXray [31]. In [34], a real-time and two-stage method based on images from a 3D laser scanner is proposed. The method performs four-class classification of narrow lap welds. Also, a comparison on classical and deep classification methods is performed with average accuracy of 80% for classical approaches while for deep methods of VGG-16 [35] and ResNet50 [36], 97.1% and 97.8% accuracy is reported, respectively. Wang et al. [37] presented a tutorial for weld defect detection based on DCNNs with implementation provided in PyTorch [38]. The paper provides a step-by-step approach for the data collection, pre-processing, and model designing, training, and testing.

Further investigation is performed for accurate localization of weld characteristics using deep methods. Hou et al. [14] designed a deep learning-based system for weld quality assessment. They used Sparse Auto Encoder (SAE) to extract and use intrinsic features for classifying 32×32 pixels weld patches and finally using a sliding window to classify image pixels as defect or non-defect. The process reaches an accuracy of 91% on GDXray [31], even though the work is a binary class defect classification and the process is time-consuming because of the nature of the sliding window approach and size of full weld images. In [39], extensive experiments with

24 various computer vision-based weld object detection methods (including deep learning methods based on sliding window) are performed and reported. In [40], two-stage detectors (i.e., Faster RCNN [56]) are used whose task is object detection of weld defects in shipbuilding which accounts for 60% of the building process, where radiography testing is used to inspect welded joints. The proposed object detector is trained to detect two general types of Porosity and Lack of Fusion/Slag defects. Moreover, the best result is acquired by data augmentation, which reached 53.2 mean Average Precision (mAP) on Faster RCNN [56] with ResNet50 [36] backbone.

Gau et al. [41] developed a contrast enhancement conditional Generative Adversarial Network (GAN) to address the contrast and class-imbalance issue. There are two separate target networks in their work. The first network accepts a 71 × 71 pixels patch from weld seam to classify the patch as defect/non-defect. For determining defect type, defective patches are passed into a second classification network. At the end, the sliding window approach is used for localizing defects. Thus, with respect to the two-stage design of the system and the sliding window, the entire system will not perform in real-time for high-resolution images. In [42], a defect localization method based on U-net and augmentation using conditional GAN (cGAN) [43] is presented, and the method is evaluated on GDXray dataset [31]. Although the method shows AUC of 88.4% for defect segmentation, lack of defect classification is discernible. Gantala and Balasubramaniam [44] presented an Automatic Defect Recognition model trained on Total Focusing Method (TFM) imaging dataset and Finite Element simulated dataset with addition of noise and further expansion of dataset utilizing Deep Convolutional GAN (DCGAN). Their two-class defect detection model were evaluated with yolov4 [45] and reached 85 Average Precision (AP) on the noisy dataset.

Although the above research papers are mostly related to employing deep CNN methods to automate the preliminary inspection in construction and welding, studies using deep CNN methods for NDT and defect diagnosis are not limited to radiography images and weld construction. Y. Yan et al. [46] developed deep models for enhanced feature extraction and ultrasonic pattern recognition for inspection gas pipelines. The method uses contact-less dual-mode bulk wave Electromagnetic Acoustic Transduce (EMAT) and interpretations of A-scan signals to detect defects. It leverages Continuous Wavelet Transform (CWT) to extract frequency-time domain features, then a deep CNN model is applied to perform high-end feature extraction, and finally, a pre-trained SVM is used for defect/non-defect classification of signals. The method feature extraction ability is verified by comparing to other methods, including Discrete Wavelet Transform (DWT) and Statistical Features, all of which are outperformed by the CNN model, which achieves 93.75% accuracy on a dataset of pipe with artificially manufactured defects. The work is performed for defect/non-defect classification, and the possibility of defect type classification is to be investigated.

In addition to ultrasonic pattern recognition, deep CNNs are also utilized for thermography crack detection. In [47], J. Hu et al. explored supervised thermography video sequence metal crack detection and localization. The work uses Eddy

Current Pulsed Thermography (ECPT), a multi-physics coupling method, to detect turbulence in conductive materials by analyzing thermal patterns. Initially, Principal Component Analysis (PCA) is used to extract thermal sequence components from original data. Then, Faster-RCNN [56] is used to perform object detection on images accurately. Finally, the method is compared to traditional detection methods, and it demonstrates 0.97 Probability of Detection, which outperforms the accurate prior method by 26%. Proposed methods are validated experimentally and have shown significant improvement in their own type of NDT and data acquisition, demonstrating the advantages of using CNN for feature extraction in NDT. While UT and Thermography methods (e.g., ECPT) are commonly used for in-line inspection and maintenance purposes and not for weld construction inspections, These methods have their limitations, such as low sensitivity to small defects or internal crack detection [9].

Studies mentioned above are all experimentally evaluated on either (1) A set of images are from a private dataset (i.e., usually created for experimental purposes) (2) GDXray [31] or similar public, and non-comprehensive sets. As shown in Figs. 6.2 and 6.3, there are noticeable differences in images from welded-joints at Stanley, and the GDXray dataset. First, GDXray has a limited number of samples. Second, class diversity is limited and also annotations and weld characteristics are based on a different standard [31]. Third, visibility of defects are limited comparing to defects at Stanley. Also, in some cases, a single patch contains more than one type of defect, which does not permit the experts to designate a single label for the entire image patch. All of which makes classification only or defect/non-defect localization incompatible with real-world industrial requirements and standards. In other words, the detection of non-hand-picked and diverse real-world samples is more of a challenge. On the other hand, since the systems will work as assistant to NDT experts, and there are limitations in hardware for deploying as well as time-constraint processing requirements, scalability is required for efficient and optimized utilization. Considering mentioned reasons, these methods either fail to reach required specifications or do not meet required performance based on industry measures.

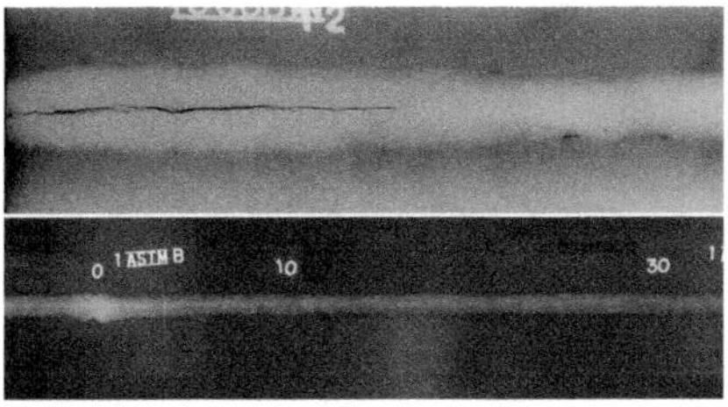

Figure 6.2 Two samples of images in the GDXray database (Top) and SBD dataset (Bottom). The bottom image is cropped to be able to compare with each other. Defects are more visible in the GDXray database than SBD dataset.

Figure 6.3 An image sample from SBD database in that two surfaces with different thicknesses is welded.

This chapter aims to address the accuracy and inference time trade-offs by presenting an efficient and scalable set of deep models. Moreover, instead of assigning a single label for each patch, accurate location and label for each discontinuity will be determined. The contributions in this work are as follow (1) describing an efficient and scalable system for object detection or classification of weld characteristics on long, high-resolution radiography weld images, which is deployable as a real-time assistant for NDT experts, (2) demonstration and analysis of the transferring augmentation strategies during training which can improve the performance of the system on detection of rare small discontinuity which are easier to miss during manual inspection and harder to detect with deep learning methods, (3) analyze and experiment with different components of the deep model, such as activation functions, and feature extraction backbones, and (4) comparative analysis on the presented models with base-line models.

The rest of this article is organized as follows: In sections 6.2 and 6.3, an overview of dataset preparation and proposed methods are provided, respectively, as well as descibtion of system architecture. In section 6.4, the methods are tested, various models are described, and the augmentation approach and results are evaluated. Finally, conclusions are proposed in section 6.5.

6.2 DATASET

The Dataset contains thousands of X-ray images taken with the purpose of NDT of weld construction in preliminary stages. There is little to no material variation in weld construction, which helps developing a model focusing on accuracy and robustness. The majority of the structures are plain carbon steel. The diameter of the pipes range from 24 to 56 inches. However, pipes with either 36 inches or 42 inches are mostly common. Moreover, the pipes Wall thickness is at least 0.5 inches with the grade of X65 or greater. Finally, all pipes are consistent with API 5L [48] in terms of types, dimensions, material, and grade.

Welded-joint images have various resolutions depending on the exterior diameter of the structure. In this dataset, the resolution of the images is roughly 15360 × 1024 pixels, with the occurrence of weld discontinuities. As the welded area only covers one-fifth of each weld image's center area, images are cropped into 224 × 224 patches with 20% overlap. This overlap benefits in two ways. First, it assists to retain defects lying in between two patches in one patch. Second, as smaller defects shift in two consecutive images, it can be interpreted as data augmentation. Next, experts annotated the images based on API 1104 [49] standards. Most of the defect-free patches are removed from the dataset to prevent overwhelming the network with non-defect images. Finally, Fig. 6.4 shows samples of the dataset, and Table 6.1 shows the distribution of images for each set. As the dataset reveals, about 75% is used as train set(i.e., 17872 images), 10% and 15% are used as dev/validation set and test set, respectively. Note that the dataset is collected from welding of various structures and different welding devices. Thus, results obtaining from this dataset can demonstrate the generalizability and robustness of proposed solutions for extensive

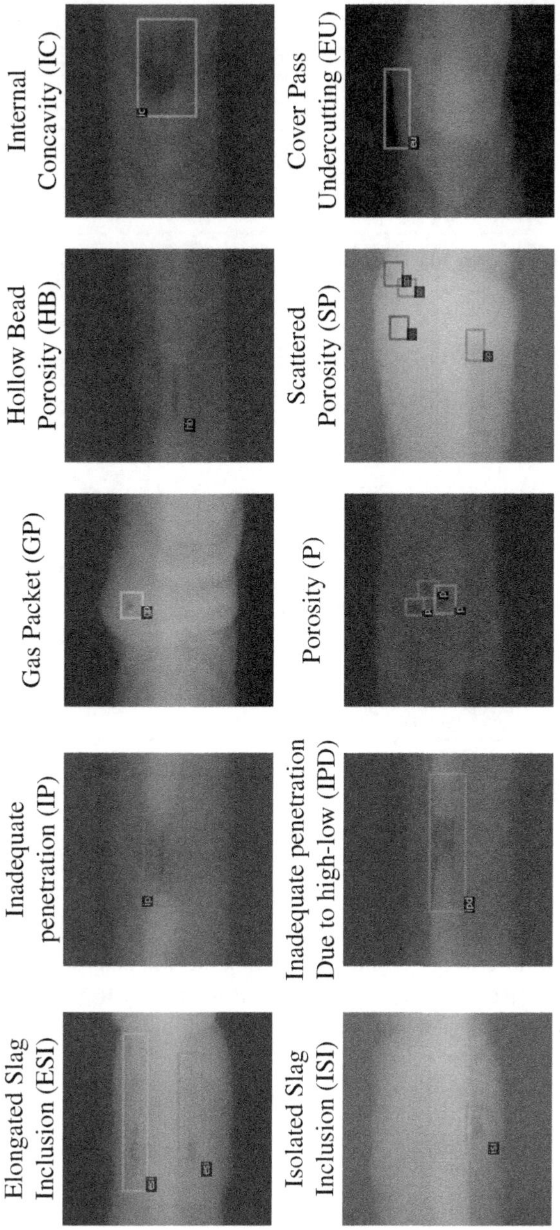

Figure 6.4 Samples of Classes in Dataset.

Table 6.1
Distribution of images and labels.

	Total	Categories									
		ESI	ISI	IP	IPD	GP	P	HB	SP	IC	EU
Train	17872	4424	2948	908	702	1840	598	4062	498	597	1295
Validation	2203	569	338	118	97	280	71	435	51	87	157
Test	3394	891	490	190	139	408	116	707	83	130	240
Total	23469	5884	3776	1216	938	2528	785	5204	632	814	1692

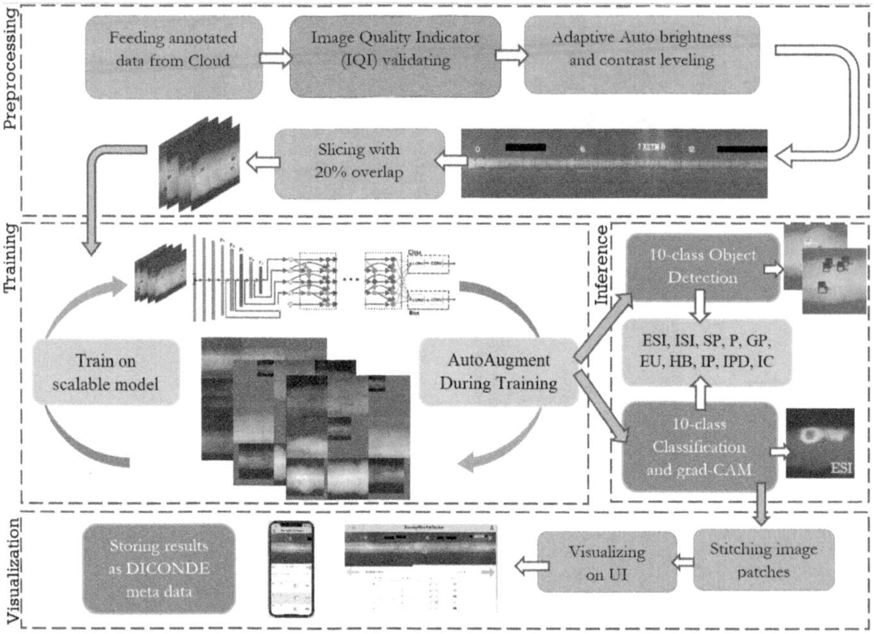

Figure 6.5 An overview of the system architecture: The box in blue accent is optional and can be done once images are captured before the start of the detection process. The arrows in gray color show transitions to the next stage. In section Inference, one of the object detection or classification tasks will be done based on the in-use trained model. In Visualization, depending on the need, stitching, visualization, storing, or all of them in once can be done.

use as assistant to NDT experts. Fig. 6.5 summarizes pre-processing steps on the dataset. The steps are described in details in section 6.3.1.

6.3 METHOD

Addressing robustness, accuracy, and time performance are required for employing a deep convolutional model in production for the task of weld defect object detection. Over recent years, scaling up image resolution, depth and width of the network, and using a larger backbone are widely used to boost the performance of the models [50, 51, 52, 53]. However, this costs in a larger model with higher computation and inference time [50, 51] as well as longer training time. Thus, a robust and efficient design is required to address the accuracy versus time performance. In order to address the trade-off between accuracy and time and achieve efficiency in models, a family of one-stage and scalable models called EfficientDet [51] are exploited. Employing a single compound coefficient, one can scale the architecture to address the trade-off between model size and accuracy of the model, resulting in a model deployable on various end-devices ranging from mobile devices to High-Performance GPU Clusters.

A two-stage object detection model generally starts with a search on regions of interest (ROI) using the selective search or, in more recent designs, applying Region Proposal Networks (RPNs), and then by passing image to the second stage for feature extraction, classification of the boxes, and refinement of the bounding box are performed [56, 57]. Although the tow-stage methods might lead to higher accuracy, the inference time because of the first stage burden is significant in sight of the additional step (RPN). In contrast, one-stage detectors apply a feature extractor called backbone and then fuse multi-level extracted features. In the end, class/box networks help to extract class labels and regression of bounding boxes. Since the image passes only once through the network, the one-stage detector performs significantly faster than other methods [53]. By utilizing pre-trained backbones, the power from classification tasks transfers to these object detectors as employed in [54]. In this section, Preprocessing steps, EfficientDet architecture design and Augmentation Strategies for object detection as well as system architecture to achieve an accurate model with low time latency are discussed, respectively.

6.3.1 SYSTEM ARCHITECTURE

Figure 6.5 depicts the required pre-processing steps to generate the dataset, which start with downloading image patches and quality validation. Although the images on the cloud storage are pre-validated for quality, it can be done through a wire IQI tag, which is discernible on the image in Fig. 6.6. As this step is optional and can be done upon uploading the images to the cloud storage, its time burden is disregarded from total system time performance. As the final two steps, brightness correction and contrast leveling, as well as slicing of the original 15360×1024 pixels image with 20% overlap is done.

As Fig. 6.5 -Training depicts, training starts on a scaled model, which depends on a single coefficient for the determination of depth and width of the network. In addition, Autoaugment during training is performed. Procedures of scaling, augmentation and network design are elaborated in 6.3.2, 6.3.3, and 6.3.4. As the next step, based on the type of the trained network in the model predicts either label and accurate location of the defects or assigns a single label for the whole patch with explanations on the decision provided. Finally, Fig. 6.5 -Visualization indicates stitching as the first step of visualization. Since exact slicing points are saved during slicing, relative

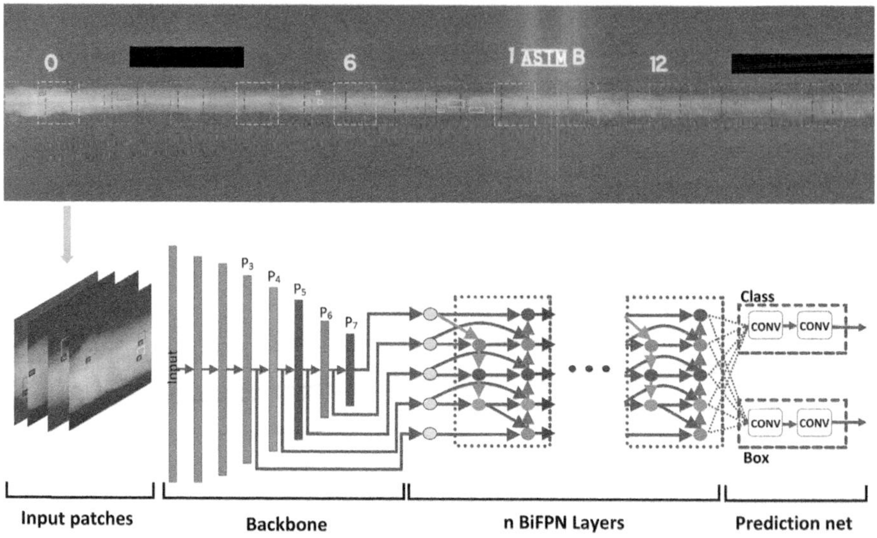

Figure 6.6 EfficientDet family architecture: Images pass through the backbone, and feature scales P3 to P7 get fed into the BiFPN network. Input image resolution is calculated from Eq. (6.1). The number of BiFPN repeated blocks extracted using Eq. (6.2). Depth of box/class prediction nets is determined using Eq. (6.3).

predicted defect locations of the whole image are calculable. Finally, the full DICONDE image can be visualized through Stanley web-app or mobile-app or saved as DICONDE metadata.

6.3.2 NETWORK

As described in section 6.2, the final dataset contains 23469 image patches of size 224 × 244 pixels. An image patch passes through backbone for feature extraction. In this work, EfficientNet [52] is used as the backbone of object detection models and for feature extraction in classification models. However, for weld quality assessment, different backbone performances are evaluated. Fig. 6.11 shows class activation maps for extracted features through several backbones. Next, multi-scale features from level P_3 to P_7 pass through a successor of Feature Pyramid Networks (FPNs) [55]. P_i denotes the resolution of the input activation map that is $1/2^i$ of the original input image. In Conventional FPN, it is assumed that features from various scales contribute equally to the final detection. A few works have investigated the optimization of feature fusion, e.g., NAS-FPN [50] is an effort to find optimum architecture for cross-scale fusing network through search. However, it takes thousands of GPU hours to find an optimal design and the resulting model is oversized. To address the equal contribution of different scales in fusing features, EfficientDet uses bidirectional FPN (BiFPN). In BiFPN, similar to FPN, a top-down pass is used, and similar to PA-Net [58] bottom-up pass is added. Nonetheless, the bottom-up pass adds a lot of costly additional weights to the network. Thus, nodes with single connections (highest and lowest levels) are removed in view of less contribution in feature fusion to optimize the structure. In addition, a few edges from input to output (similar to skip connections in ResNet [36]) are added, which boost both the training and accuracy processes. Finally, fused features pass through two similar class and box networks used to determine the class label and the bounding box location of detected discontinuities. Similar to backbone and BiFPN, depth of class/box nets get scaled with a single coefficient.

6.3.3 SCALABILITY

In this part, the single compound scaling coefficient of the overall architecture is reviewed. EfficientDet family starts from the smallest model D0 and ends with the deepest and largest model D7, where the number stands for the single compound coefficient ϕ, used to scale input image resolution and overall depth and width of the architecture. For backbone, if EfficientNet is used, one of the pre-trained networks is applied based on ϕ. Fig. 6.6 shows the architecture, which is similar for all networks. The final input image resolution gets determined using Equation (6.1).

$$R_{\text{input}} = 512 + \phi \cdot 128 \tag{6.1}$$

Equation (6.2) shows how the number of channels/layers of the BiFPN is scaled, where each layer is one of the BiFPN repeated blocks starting from 3 for D0 shown in Fig. 6.6. Finally, the number of layers of the class/box gets determined through Equation (6.3).

$$W_{\text{bifpn}} = 64 \cdot (1.35^{\phi}), D_{\text{bifpn}} = 3 + \phi \tag{6.2}$$

$$D_{\text{box}} = D_{\text{class}} = 3 + \lfloor \frac{\phi}{3} \rfloor \tag{6.3}$$

6.3.4 DATA AUGMENTATION

Many object detection as well as weld quality assessment deep learning approaches employ data augmentation in order to improve both the performance of the network and generalization [32, 42, 40]. The effectiveness of augmentation is shown and evaluated in literature [62]. Nonetheless, there are countless strategies, such as rotation, affine, zoom in/out, flipping, etc., various magnitudes, and also different possible combinations of strategies to be used for augmenting the dataset. One solution is to search through all possible solutions to find the optimal ones. Authors in [60] investigated and searched through the area of 10^{10} different combinations for the classification task. Similarly, [61] investigated the effectiveness of AutoAugmentation for object detection and extracted a few sets of policies enhancing detection performance the best for the object detection task named as policy V0-3. As searching for optimal augmentation strategies is a time-consuming task, extracted policies are applied and investigated in this work. For this purpose, a base model (D0 with EfficientNet B0 backbone) is trained utilizing each of the policies to find the best policy. Then best policy is used for training larger models and investigating other effective parameters of the model.

6.3.5 EVALUATING METRICS

Evaluating results is performed through Average Precision (AP) metrics. Models output a bounding box, a corresponding class label, and confidence for each detection. A detection is considered correct when the area of the ground truth bounding box and the detected box have at least 0.5 intersection over the area of the union of two mentioned boxes, which is called Intersection over Union (IoU). Also, the class labels of both bounding boxes should be the same, which means:

$$IoU = \frac{area(B_{\mathrm{Bp}} \cap B_{\mathrm{gt}})}{area(B_{\mathrm{Bp}} \cup B_{\mathrm{gt}})}. \tag{6.4}$$

With IoU less than 0.5, the detection is counted as fp. Fn is also the count of non-detected bounding boxes. Therefore, precision and recall are calculated through the following :

$$precision = \frac{tp}{tp + fp}; recall = \frac{tp}{tp + fn} \tag{6.5}$$

As recall and precision of a robust object detector do not alter much with varying confidence, it is required to consider multiple confidence thresholds to evaluate the performance of the object detector [75]. Defining all-point interpolation of the area under precision-recall curve obtains accurate results by pruning zig-zag behavior of the curve and utilizing maximum precision ($P_{\mathrm{interp}}(r)$ where r is recall level and recall of the point is greater than r_{n+1}) at each recall level, instead of using the precision at that point. The mathematical presentation of this is as follows:

$$AP = \sum_{n} (r_{n+1} - r_n) P_{\mathrm{interp}}(r_{n+1}) \tag{6.6}$$

Where,

$$P_{\mathrm{interp}}(r_{n+1}) = \max_{\tilde{r}:\tilde{r} \geq r_{n+1}} P(\tilde{r}).$$

AP has become a standard for comparing model performance in different object detection challenges [63] as well as literature [59, 51, 56, 64].

In section 6.4, models are evaluated using mAP (mean AP)(which is equal to mean of AP with IOU threshold ranging from 0.50 to 0.95 and step of 0.05), AP_{50}, AP_{75} (which is equal to AP@IoU=0.75), AP_s (s stands for small and objects with area $< 32^2$) , AP_m (m is medium and area of the objects are between 32^2 and 96^2), as well as AP_l (objects with area $> 96^2$).

6.4 EXPERIMENTS

In this section, various experiments are designed and performed to investigate a set of scalable models with fast processing time while maintaining high accuracy. In addition to EfficientNet backbones, results are reported utilizing other backbones namely, MobileNetV3 [65], Resnet50 [36], which is called Resdet50 in detection models, CspResdet50 [66], and Darknet (utilized in Yolov3 [53]). Moreover, stanalone object detection models including Yolov3, Yolov4 [45], Yolov5 [80], and RetinaNet [64] are fine-tuned as a basis for comparison. In the following subsections, the K-means method is used to extract optimal anchor boxes, analysis and results from applying various AutoAugment policies on models, training setup and hyperparameter tuning, quality assessment with single class labels, effects of using several activation functions, and backbones are elaborated, respectively.

6.4.1 ANCHOR BOXES

Similar to [55], EfficientDet uses anchor boxes to detect objects. By default there are three distinct aspect ratios (0.5,1.0,2.0). K-means clustering is utilized to find the set of optimal aspect ratios for the box prediction network [59]. Moreover, the input image resolution is also considered in optimized aspect ratio calculation. Table 6.2 demonstrates the effectiveness of using aspect ratios calculated by K-means. The results are reported using AP metrics. New aspect ratios (1.2, 2.14, 3.8) suggest that 99.92% (i.e., equivalent to the percentage of bounding boxes that lie into one of K-means calculated clusters) of the defects are horizontal rectangles, and optimizing helps with a 6.6% increase in AP_{50}.

6.4.2 ANALYSIS OF AUGMENTATION POLICIES

Since training each model employing all policies V0 to V3 is time-consuming, EfficientDet-D0 is considered the base model for analyzing how transferring augmentation policies affect the detection of characteristics. Table 6.3 demonstrates the effects of utilizing various policies during training for augmentation, based on AP metrics. In NoAugment, raw images are passed to the network, while In Train-timeAug, two common augmentations for train-time are used. First, images are flipped horizontally with a probability of 50%, and second, images are randomly

Table 6.2

Effect of optimizing aspect ratio of anchor boxes based on bounding box annotations in train dataset. For this experiment EfficientDet-D0 model is used and the optimized aspect ratios are (1.2, 2.14, 3.8).

default					*k-means optimized* (% of improvement)				
mAP	AP_{50}	AP_s	AP_m	AP_l	mAP	AP_{50}	AP_s	AP_m	AP_l
30.7	59.3	24.0	33.9	68.3	34.8(↑**4.1**)	65.9(↑**6.6**)	25.4(↑**1.4**)	35.0(↑**1.1**)	69.2(↑**0.9**)

Table 6.3
Policies used to train on D0 model on entire train images. Results are reported on the validation set.

policy Name	Base (% of Improvement)			
	mAP	AP_s	AP_m	AP_{50}
NoAugment	24.2	21.5	27.2	46.0
Train-timeAug	30.4(↑**6.2**)	24.4(↑**2.9**)	39.0(↑**11.8**)	62.5(↑**16.5**)
policyV0	28.0(↑**3.8**)	23.2(↑**1.7**)	33.0(↑**5.8**)	56.0(↑**10.**)
policyV1	30.3(↑**6.1**)	22.2(↑**0.7**)	35.1(↑**7.9**)	61.1(↑**15.1**)
policyV2	30.2(↑**6.0**)	22.1(↑**0.6**)	36.3(↑**9.1**)	59.7(↑**13.7**)
policyV3	31.1(↑**6.9**)	24.9(↑**3.4**)	37.2(↑**10.**)	60.9(↑**14.9**)

resized and padded with a random scale between 0.1 to 2.0. Also, bi-linear interpolation is used while resizing, and the mean of the dataset is applied for padding when the final image is smaller than 512×512 pixels (as 512×512 pixels image is the target image size for the D0 model). PolicyV0-3, refer to each of 4 policies introduced in [61]. In a similar way, during training of the D0 model using each of these policies, a random set of strategies from the selected policy with a probability of 66% is selected, and the input image is augmented based on it (the probability of not performing any of the strategies is one-third). Moreover, similar augmentation is performed on bounding boxes if any is affected. Based on Table 6.3, augmentation policies dramatically boost the performance of the network by 3.8 to 6.9 AP. Most policies assists the network detect smaller defects (i.e., AP_s which is more of importance since are easier to miss during manual inspection and are harder to detect with deep learning methods). For further investigations, train-timeAug, and policyV3 are applied to the images as they resulted best in these experiments. Fig. 6.7 depicts a sample training batch with mentioned augmentations applied.

6.4.3 TRAINING AND HYPER PARAMETER TUNING

The size of the models and the resolution of input images increase from D0 to D7 gradually using Eq. 6.2, 6.1, and 6.3. It is not possible to train all the models on GPUs with 16GB RAM with suitable possible batch size relative to model size. Models with smaller ϕ coefficients (i.e., D0 to D2) are trained on 3 NVIDIA V100 16GB RAM GPUs with maximum possible batch size though for fitting these models in such GPU memory, a few actions are performed. First, mixed-precision training[1] is

[1]This tool is provided in the Apex package, and it is possible to apply this using a few code lines on any PyTorch network implementations. available at https://nvidia.github.io/apex

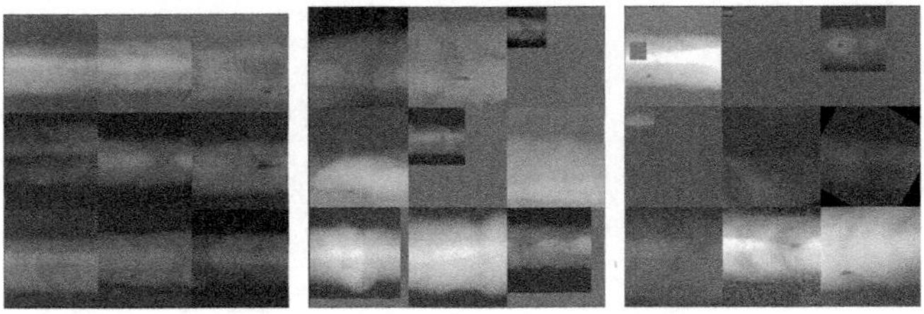

(a) (b) (c)

Figure 6.7 Samples of a training batch with different augmentations: 6.7a - a batch with no augmentation, 6.7b - an augmented batch with a collection of random augmentation named as Train-time augmentation 6.7c - augmented batch based on policy-v3.

applied using the Apex package which assists decrease memory usage and training time by utilizing half-precision weights and operations if possible. Second, as providing accurate statistics for batch normalization is crucial for the stabilized learning process and high-speed convergence, in distributed training, Synchronized batch normalization is used to provide cross-device batch-norm statistics. Nonetheless, these would not help to fit D5 to D7 models in GPU. Thus, results related to those models are not reported. Finally, for comparison, several original Yolo and RetinaNet models are trained. For RetinaNet models, images are resized to 800 $\times$ 800 pixels, and resnet with 50 or 101 layers are used as the backbone, and for Yolo models, images are resized to 640 $\times$ 640 pixels. Implementation for yolo models can be found in [80], and RetinaNet can be found in detectron2 framework [79].

During training, normalization using pre-computed mean and standard deviation values per channel on the entire dataset is performed. Also, each image is first randomly flipped horizontally and/or resized for all experiments (i.e., Train-timeAug explained in 6.3.4). For weight initialization, the weights originally were trained on MS COCO dataset in [51] and are converted to PyTorch in [67], are used. Thus, all weights of the network are trained to reach maximum performance similar to our previous work [54].

Identical to [51] Cosine Learning [68] is used. At the beginning of the training process, for the first few epochs (epoch number 0 to 5) learning rate increases gradually to the desired point, and from epoch 5 to the end of the training process the learning rate decreases gradually in cosine form. In addition, learning rate noises applied to 30% and 90% of the training process. Moreover, in a few experiments, Exponential Moving Average (EMA) [69] with a weight decay of 0.9998 was applied; however, it was removed as non-EMA training ended up with higher AP. Furthermore, a few optimizers are evaluated and results show, in this task Fusedadam [70] optimizer converges faster and reaches to a higher accuracy (0.6 mAP). In the following, the impact of different activation functions is discussed.

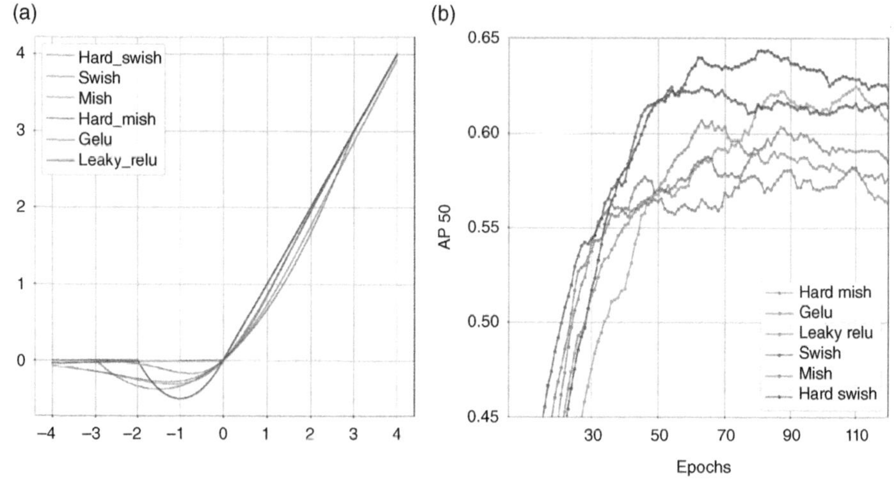

Figure 6.8 Activation Functions used for comparing performance of the models.

6.4.4 EFFECT OF DIFFERENT ACTIVATION FUNCTIONS

The performance of the base model (i.e., EfficientDet-D0) is analyzed by testing over different activation functions, namely Leaky Rectified Linear Unit (Leaky ReLU), Gaussian Error Linear Units (GeLU) [71], Swish [72], Mish [73], hard Swish [65] in which sigmoid is replaced with $\frac{relu6(x+3)}{6}$, which is more memory efficient, and hard Mish [74]. Fig. 6.8a visualize mentioned activation functions. Note that the specified activation function is used for BiFPN layers and class/box prediction nets. As shown in Fig. 6.8b, both hard Swish and Swish outperforms other activation functions based on AP_{50}. The same improvement applies for other AP metrics. However, this did not happen for deeper models of the EfficientDet family. Thus, default Swish is used to maximize model performance, though in view of the memory efficiency of hard Swish, it is a preferable choice for activation function if the model is planned to get deployed on hardware-constraint end-device. For non-EfficientDet family models, the default activation function of the model is used.

6.4.5 DEFECT OBJECT DETECTION WITHOUT CONSIDERING CLASS LABELS

In this experiment, all discontinuities are considered with a single *defect* label. Table 6.4 shows network performance considering a single class for all discontinuities. Although these models only perform localization of the defects and no class label is available, higher accuracy in localization is reached. In addition to EfficientDet Family, several other models are also added for sake of comparison. All models are trained using base parameters in their original work, except for Yolov3, in which focal loss is preferred to address class imbalance. In section 6.4.6, models in Table 6.4 are discussed.

Table 6.4

Model defect detection performance on test set based on defect/non-defect labels. Inference Times are reported based on tests on a V100 GPU card. For EfficientDet models, Number D# stands for ϕ coefficient of the model, and depth and input image resolution can be acquired using Eq. 6.1, 6.2, and 6.3. Also, the number stands for backbone number in [52]. For other models same structure of EfficientDet-D1 is used except the backbone, the name of which is determined in Model Name. For models without efficient in their name, implementations from [79] or [80] with common augmentation is employed.

Model Name	# Params	AP	AP_{50}	AP_{75}	AP_s	AP_m	Inference Time		
							b=1	b=8	b=16
Yolo v3	61.4M	34.3	81.0	-	-	-	54	65	75
Yolo v4	64.3M	27.4	61.6	-	-	-	48	130	185
Yolo v5x	87M	34.8	81.9	-	-	-	35	50	56
RetinaNet-50	37.9M	29.6	73.6	14.5	29.5	30.5	22	-	-
RetinaNet-101	56.9M	30.9	75.2	14.2	29.9	31.5	20	-	-
EfficientDet-D0	3.8M	37.1	87.5	18.5	37.5	37.3	35	130	185
EfficientDet-D3	11.9M	37.7	88.1	19.7	36.7	38.8	14	56	65
Efficientdarkdet53	11.9M	36.6	85.7	18.9	38.4	35.5	15	34	37
EfficientCspResDet50	23.6M	37.6	86.2	20.2	38.8	36.8	15	36	39
EfficientResDet50	26.9M	38.2	89.0	21.8	37.8	40.9	27	143	168

6.4.6 EVALUATING RESULTS

Tables 6.4, 6.5, and 6.6 demonstrate models evaluation on 10-class object detection task, 10-class classification, and defect/non-defect object detection task, respectively. All results for object detection models are obtained from the test set. For all experiments, common train-time augmentations and policyV3 (described in 6.3.4) are

Table 6.5

10-Class classification backbone performances.

Model Name	Image Size	# of Params	Top-1 Accuracy	Precision	Recall	F1-score	Inference Time (images/sec)	
							b=1	b=16
Resnet50	224	23.5M	86.83	85.68	86.26	85.97	30	207
CspResnet50	224	20.6M	85.3	83.89	83.71	83.8	31	266
MobileNetV3	224	4.2M	79.91	79.41	78.67	79.04	37	267
EfficientNet-b0	224	3.8M	82.15	81.4	81.36	81.38	38	355
EfficientNet-b1	240	6.5M	85.4	83.8	84.4	84.1	29	200
EfficientNet-b6	528	40.7M	85.71	84.5	84.1	84.3	13	26
EfficientNet-b8	672	89.6M	90.2	89.5	88.72	89.11	6	10

Table 6.6

10-class defect detection models performance on test set based and inference time with multiple batch sizes. For EfficientDet models, Number D# stands for ϕ coefficient of the model and depth and input image resolution can be acquired using Eq. 6.1, 6.2, and 6.3. Also, the number stands for backbone number in [52]. For other models same structure of EfficientDet-D1 is used except the backbone. Note that all models are optimized for maximum AP.

Model Name	# of Params	Test Set					Inference Time		
		AP	AP_{50}	AP_{75}	AP_s	AP_m	b^b=1	b=8	b=16
YOLO V3	61.5M	30.9	58.4	-	-	-	90	150	150
YOLO V4	64.3M	24.1	46.0	-	-	-	50	120	125
YOLO V5x	87.2M	30.1	58.1	-	-	-	35	50	56
RetinaNet50	37.9M	26.7	51.0	18.4	20.6	27.6	21	-	-
RetinaNet101	56.9M	27.7	54.7	18.2	21.6	30.7	20	-	-
EfficientDet-D0	3.8M	34.8	65.9	22.2	25.4	35.0	35	144	224
EfficientDet-D1	6.5M	34.9	68.0	23.7	27.3	34.4	17	111	148
EfficientDet-D2	8M	35.2	68.0	23.0	26.0	36.5	16	88	106
EfficientDet-D3	11.9M	34.7	69.3	25.1	25.9	34.9	13	58	66
Efficientdarkdet53	11.9M	34.4	67.1	25.4	26.4	33.9	14	33	37
EffcientCspResdet50	23.7M	34.1	68.1	22.5	25.7	33.6	13	35	39
EfficientResDet50	27M	36.1	72.4	25.0	26.0	37.7	24	132	150

b denotes batch size for inference.

applied. Although hard Swish improved accuracy for shallower models, the same did not happen for deeper ones (D3 and deeper). Thus, default activation functions of the models are used (i.e., Swish for EfficientDet models, leaky-ReLU for CspResdet50 and darkdet53, and ReLU for Resdet50). As mentioned in tables, generally, deeper models show more accurate performance. However, little improvement or minor deterioration in larger models is a result of having to use smaller batch size to be able to fit the model into the GPUs (i.e., batch size of 20 per GPU is used to train the D0 versus batch sizes 3 and 1 for D3 and D4 models, respectively), which accounts for inaccurate estimation of statistics of batch normalization and deteriorates training process. Since for task of weld quality assessment and indexing of weld as well as rejecting or accepting, predicting 50% of the discontinuity is acceptable, AP_{50} is used for further model comparison and analysis. AP_l is not reported because defects in welds with area greater than 96^2 pixels is under-sampled and uncommon. Therefore, it would not be a reliable measure to evaluate the performance of the models, and it is not reported.

For inference time analysis, a similar GPU that is used for training, NVIDIA V100 16GB, is exploited. Inference times in Tables 6.4 and 6.6 suggest, models are able to perform in real-time performance based on definition of real-time for object detection models [76]. As a result, the fastest and the most accurate models can infer up to 224 and 150 image patches per second with a batch-size of 16, respectively. Considering the fact that in the worst case, each complete weld image has a length of 15360 pixels and is cropped with 20% overlap, a full image will have about 86 patches, meaning models can infer an image in 385 to 465 ms. Thus, models are able to process weld images in real-time with consideration of required pre-processing. Fig. 6.9 summarizes models' latency and Floating-point Operations(FLOPs) count. Yolo models have a higher number of operations, and the resulting models are larger. In contrast, the EfficientDet family models and models with Bi-FPN Layers enhanced with AutoAugmentation are both smaller and more accurate. Although EfficientDet models have a smaller number of parameters, they perform slower on GPU because of slower execution of separable convolution. Finally, a fusion of Resnet50 with EfficientDet object detection architecture results in best accuracy versus latency, for this task.

In Tables 6.6 and 6.4 models reported above double line are trained for sake of comparsion. In Yolo models in addition to train-time aug, mosaic augmentation is applied. Although this improved results by 0.5 AP, EfficientDet models are still more accurate. In contrast to large number of parameters in Yolo, they still perform faster than efficientDet models and the reason is that depth-wise separable convolutions is employed for feature fusion in EfficientDet and they run slower on GPU. However, thanks to lower number parameters these model will perform better on CPUs in compare with competitors. Results from Table 6.4, in which for all discontinuities, a single label is used, suggest that a portion of false-positive detections are related to detecting correct labels. In the following, erroneous and missed detections of the best model are elaborated upon. Also, the performance of feature extraction backbones both numerically and visually is discussed.

Error Analysis: Based on Table 6.6 and inferred images of the test set, erroneous detections of the network with highest AP_{50} belong to one of these subcategories: 1) errors as a consequence of inadequate IoU of detections and ground truth: 0.7% of samples from 3 classes (i.e., IP, ESI, and GP were 76% where from class ESI) of test set belong to this category. Fig. 6.12c shows a sample from this category. 2) False-positives were mostly related to instances that a non-defect bounding box is detected, and it is closely similar to one of the other defect classes, and a non-expert observer might consider it as a defect. However, it does not meet minimum requirements such as length for Slag Inclusion, size for Porosity, and other criteria to be counted as a discontinuity. Finally, out of 4.5% of instances lying in this category, Slag Inclusions and HBs had the largest normalized percentage of errors. Mostly, sides of the weld root and also weld toe were falsely predicted as Slag Inclusions. Fig. 6.12e is a sample of this category. As a workaround to reduce the error rate of this type, adding a large number of similar image patches to the train set is suggested. 3) False-negatives, where the network does not detect defects. With more than 12% of false-negatives, this group contains the largest erroneous behaviour of the model, with HBs and ESIs forming more than 55% of the normalized number of false positive detections. Fig. 6.12f is a sample of a false-negative. A suggested workaround is to perform online or offline hard example mining for training. Note that by lowering the minimum confidence threshold, most of these are detected concisely by the

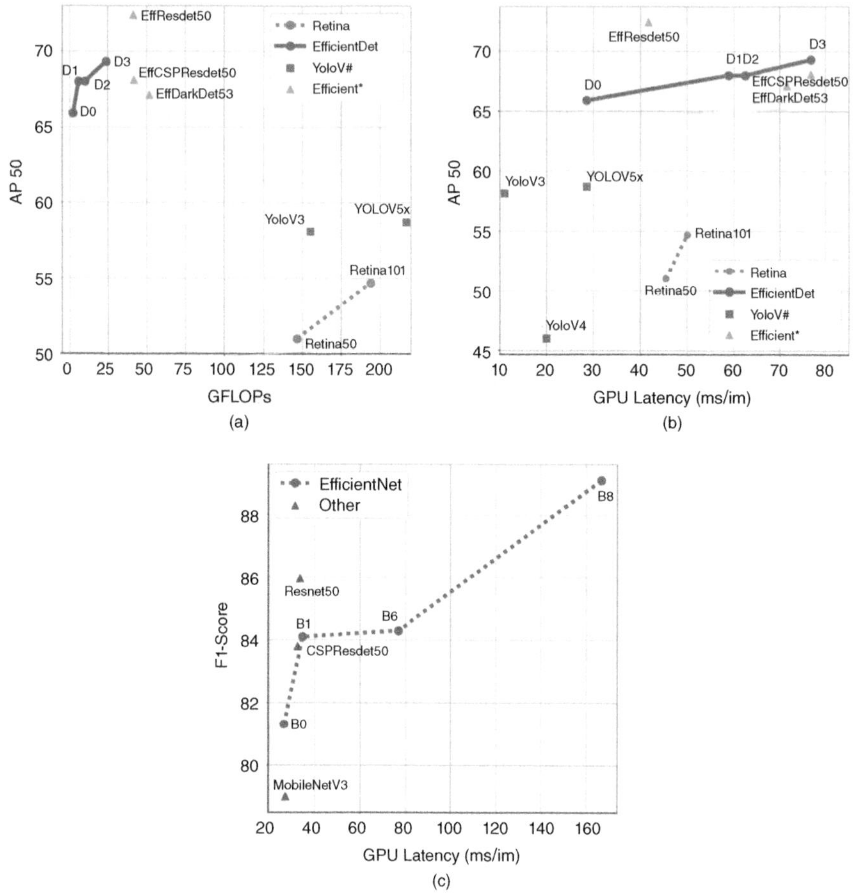

Figure 6.9 (a) - AP_{50} accuracy versus giga Floating-point Operations (GFLOPs) for 10-class object detection models. (b) shows AP_{50} versus GPU latency (milliseconds/224 $\times$ 224 image patch) for 10-class Object detection. (c) shows F1-score versus GPU latency (ms/224 $\times$ 224 image patch) performance for 10-class classification models.

network. 4) miss-classified samples, in which the network detects the object with acceptable IoU, though the class label is incorrect. A sample will be Fig. 6.12d. Finally, Fig. 6.10a shows the distribution of miss-classified samples from the 10-class object detection model. It is showing that HB class has the most miss-class detections, and it is mostly mistaken with class IC, and the similarity is that both IC and HB create a hallow area in the weld root.

Comparison of Backbones: Although it is common that multiple discontinuity types appear in a single patch, a part of the dataset (which includes around 80% of each set) that holds image patches with a single defect type is separated and used to evaluate feature extraction and backbone performance, and also to train a classification model. Table 6.5 shows performance of various backbones. The most

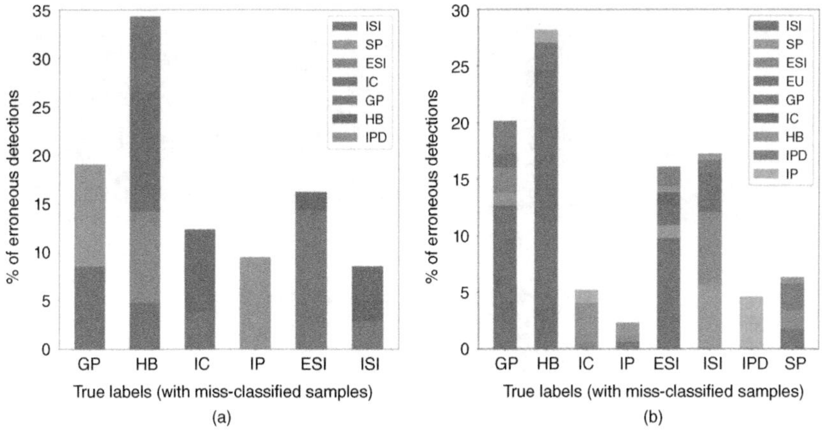

Figure 6.10 miss-classified detected defects. Miss classification distribution for 10-class object detection model–a, Number of miss-classfied patches in classification backbone (EfficientNet-b8)–b.

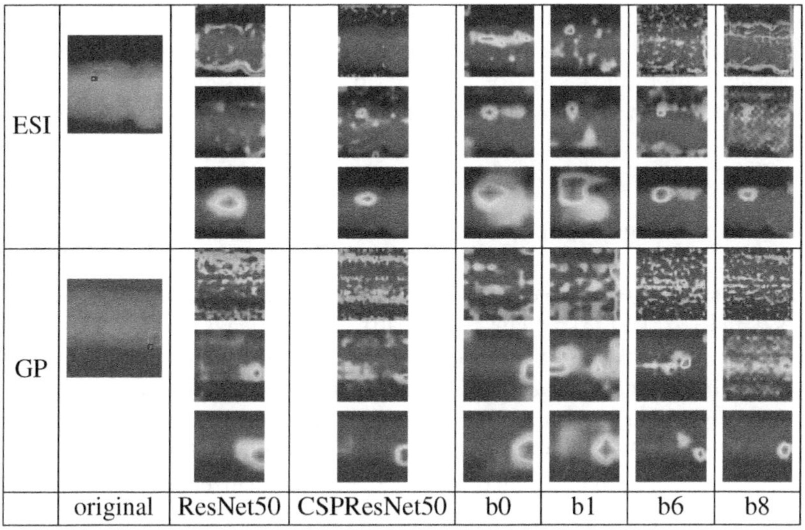

Figure 6.11 grad-CAM visualization of different backbones from their three last blocks for classes GP and ESI.

accurate backbone is EfficientNet-B8, with 90.2% accuracy on the validation set. A similar training environment, and optimizer with object detection models are used for this purpose. Transfer learning is applied and weights were originally trained on ImageNet [74]. Fig. 6.10b shows the distribution of erroneous behavior of the classifier. Based on Fig. 6.10b, most of the miss-classified samples are related to

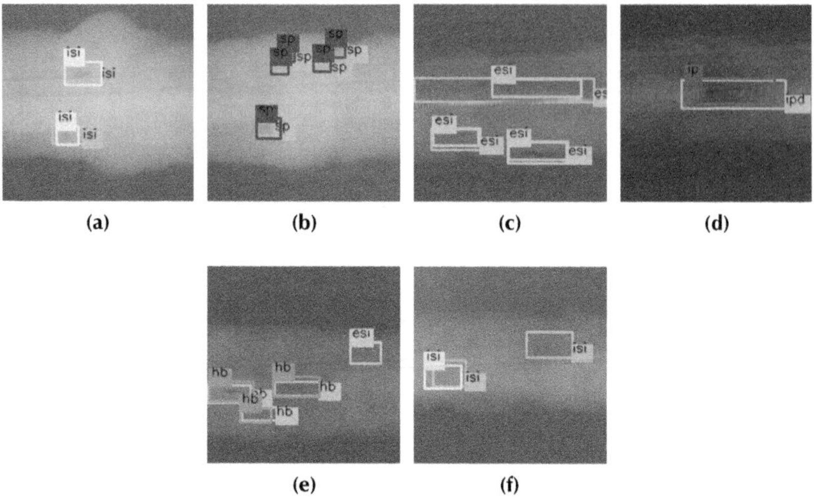

Figure 6.12 Examples of Network Prediction: True Positives - 6.12a and 6.12b (green boxes denote ground truth and others are network prediction). False Negatives and False positive as a result of IoU < 0.5 - 6.12c, erroneous class prediction because of the similarity of the classes IP and IPD - 6.12d, Hard samples which resulted in incorrect prediction as a consequence of not meeting minimum standards - 6.12e and 6.12f.

HB. Also, erroneous detections are mostly missclassified as ISI as shown in Fig. 6.10b.

Explainability using Grad-CAM: Gradient-weighted Class Activation Mappings (Grad-CAMs) [78] recognizes the parts of input image with deterministic role in final decision-making of the model. In Grad-Cam, instead of applying Global Average Pooling as ending layers [77] which requires model modification and affects the network performance, back-propagation is utilized to extract feature contributions. Therefore, class activation maps get extracted precisely. In Fig. 6.11, Grad-Cams of various backbones with different depth are visualized, which provides local explainability for input images. bottom image of each cell shows final layer grad-cam, and upper images are second to last and third from last block output. It shows how network gradually attend to discriminative features of each image.

6.5 CONCLUSIONS

In this chapter, a scalable and efficient family of deep models for 10-class weld quality assessment using object detection is presented. A comparative analysis on various models is also performed; several critical elements of the networks such as Activation Functions and hyperparameters are explored and tuned to achieve state-of-the-art results on the dataset. Moreover, the effects of transferring Object Detection

AutoAugment policies are surveyed. Furthermore, various scenarios such as considering task as a classification only task and defect/non-defect scenarios are also analyzed and models are compared with main-stream object detection models in real-time applications. Finally, model visual explainability is analyzed through employing grad-cam and visualizing gradient information for target class. The results are interpreted as The results demonstrate that models are able to infer a complete welded joint (15360 × 1024 resolution X-ray Image) in 385 milliseconds. Although classification task outperforms object detection models, localization of the defect (whether the defect is on root pass, fill pass or cover pass) is necessary for further indexing of the weld, pass or rejection, and optimization of welding operation.

Traditional computer vision techniques for weld defect detection require several critical pre-processing steps resulting in a non-robust outcome or human intervention is needed. In contrast, automatic feature extraction approaches and deep learning-based methods require minimum human intervention or pre-processing to achieve state-of-the-art results. The models presented here can be used as assistive defect-recognition systems to facilitate robust defect localization and classification and to reduce both human workload and error. Finally, as experts may have conflicting and personal performance in particular defect detection, provided deep models may train on specific samples and predict defects with a consolidated standard which can also be helpful in training experts.

Future works contain test-time augmentation, model ensemble without sacrificing real-time capability of the system, searching for optimal auto augmentation policies utilizing reinforcement learning since policies were initially extracted from the coco dataset and the nature of the weld images is not consistent with nature of coco dataset images. In addition, through time, more samples will be gathered from various sites of different parts of the world, and the dataset will expand both in the number of classes and number of instances per class.

ACKNOWLEDGEMENT

This work is done primarily with the help and support from Michael George, Jeremy Guretzki, Matthew Nelson, Jason Miller, William Aston, Jake Smith, Haresh Ghansyam, Pete Morris, Prashanth Tirumalaseti, Adam Wynne Hughes, and Shengnan Wang as well as great support from Dr. Mark Maybury and Dr. Manish Mehta from the office of CTO at Stanley Black and Decker.

REFERENCES

1. R. Vilar, J. Zapata, R. Ruiz, "An automatic system of classification of weld defects in radiographic images," *NDT & E International*, vol. 42, no. 5, pp. 467–476, 2009.
2. M-M. Naddaf-Sh, S. Naddaf-Sh, H. Zargaradeh, S. M. Zahiri, M. Dalton, G. Elpers, A. R. Kashani, "Next-Generation of Weld Quality Assessment Using Deep Learning and Digital Radiography," in *2020 AAAI Spring Symposium Series*, 2020.

3. M-M. Naddaf-Sh, S. Naddaf-Sh, H. Zargarzadeh, S. M. Zahiri, M. Dalton, G. Elpers, A. R. Kashani, "Defect detection and classification in welding using deep learning and digital radiography," in *Fault Diagnosis and Prognosis Techniques for Complex Engineering Systems*, H. Karimi, Ed. Academic Press, 2021, pp. 327–352.
4. A. Shahriar, R. Sadiq, S. Tesfamariam, "Risk analysis for oil & gas pipelines: A sustainability assessment approach using fuzzy based bow-tie analysis," *Journal of Loss Prevention in the Process Industries*, vol. 25, no. 3, pp. 505–523, 2012.
5. Z. Rui, G. Han, H. Zhang, S. Wang, H. Pu, K. Ling, "A new model to evaluate two leak points in a gas pipeline," *Journal of Natural Gas Science and Engineering*, vol. 46, pp. 491–497, 2017.
6. L. T. Popoola, A. S. Grema, G. K. Latinwo, B. Gutti, A. S. Balogun, "Corrosion problems during oil and gas production and its mitigation," *International Journal of Industrial Chemistry*, vol. 4, no. 1, pp. 1–15, 2013.
7. S. M. Anouncia, R. Saravanan, "Non-destructive testing using radiographic images a survey," *Insight-Non-Destructive Testing and Condition Monitoring*, vol. 48, no. 10, pp. 592–597, 2006.
8. X. Dong, C. J. Taylor, T. F. Cootes, "Small defect detection using convolutional neural network features and random forests," in *Lecture Notes in Computer Science*, vol. 11132, 2019, pp. 398–412.
9. Q. Ma, G. Tian, Y. Zeng, R. Li, H. Song, Z. Wang, B. Gao, K. Zeng, "Pipeline In-Line Inspection Method, Instrumentation and Data Management," *Sensors*, vol. 21, no. 11, 2021.
10. A. Shukla, H. Karki, "Application of robotics in offshore oil and gas industry— A review Part II," *Robotics and Autonomous Systems*, vol. 75, pp. 508–524, 2016.
11. L. Yang, Y. Liu, J. Peng, "Advances techniques of the structured light sensing in intelligent welding robots: a review," *The International Journal of Advanced Manufacturing Technology*, pp. 1–20, 2020.
12. S. Habibian, M. Dadvar, B. Peykari, A. Hosseini, M. H. Salehzadeh, A. H. M. Hosseini, F. Najafi, "Design and implementation of a maxi-sized mobile robot (Karo) for rescue missions," *ROBOMECH Journal*, vol. 8, no. 1, Jan. 2021.
13. M. Dadvar, S. Habibian, "Contemporary Research Trends in Response Robotics," arXiv preprint arXiv:2105.07812, 2021.
14. W. Hou, Y. Wei, J. Guo, Y. Jin, C. Zhu, "Automatic Detection of Welding Defects using Deep Neural Network," *Journal of Physics: Conference Series*, vol. 933, no. 1, 2018.
15. X. Dong, C. J. Taylor, T. F. Cootes, "A Random Forest-Based Automatic Inspection System for Aerospace Welds in X-Ray Images," *IEEE Transactions on Automation Science and Engineering*, pp. 1–14, 2020.
16. D. Mery, M. A. Berti, "Automatic detection of welding defects using texture features," *Insight-Non-Destructive Testing and Condition Monitoring*, vol. 45, no. 10, pp. 676–681, 2003.
17. J. Kumar, R. S. Anand, S. P. Srivastava, "Multi-class welding flaws classification using texture feature for radiographic images," in *2014 International Conference on Advances in Electrical Engineering (ICAEE)*, 2014, pp. 1–4.
18. J. Kumar, R. S. Anand, S. P. Srivastava, "Flaws classification using ANN for radiographic weld images," in *2014 International Conference on Signal Processing and Integrated Networks (SPIN)*, 2014, pp. 145–150.

19. J. Hassan, A. M. Awan, A. Jalil, "Welding Defect Detection and Classification Using Geometric Features," in *2012 10th International Conference on Frontiers of Information Technology*, 2012, pp. 139–144.
20. O. Zahran, H. Kasban, M. El-Kordy, F. E. Abd El-Samie, "Automatic weld defect identification from radiographic images," *Ndt & E International*, vol. 57, pp. 26–35, 2013.
21. T. Y. Lim, M. M. Ratnam, M. A. Khalid, "Automatic classification of weld defects using simulated data and an MLP neural network," *Insight-Non-Destructive Testing and Condition Monitoring*, vol. 49, no. 3, pp. 154–159, 2007.
22. J. Zapata, R. Vilar, R. Ruiz, "Performance evaluation of an automatic inspection system of weld defects in radiographic images based on neuro-classifiers," *Expert Systems with Applications*, vol. 38, no. 7, pp. 8812–8824, 2011.
23. I. Valavanis, D. Kosmopoulos, "Multiclass defect detection and classification in weld radiographic images using geometric and texture features," *Expert Systems with Applications*, vol. 37, no. 12, pp. 7606–7614, 2010.
24. W. Hou, D. Zhang, Y. Wei, J. Guo, X. Zhang, "Review on computer aided weld defect detection from radiography images," *Applied Sciences*, vol. 10, no. 5, p. 1878, 2020.
25. M. A. Carrasco, D. Mery, "Segmentation of welding defects using a robust algorithm," *Materials Evaluation*, vol. 62, no. 11, pp. 1142–1147, 2004.
26. D. Mery, "Automated detection of welding discontinuities without segmentation," *Materials Evaluation*, vol. 69, no. 6, pp. 656–663, 2011.
27. M. Ben Gharsallah, E. Ben Braiek, "Weld inspection based on radiography image segmentation with level set active contour guided off-center saliency map," *Advances in Materials Science and Engineering*, vol. 2015, 2015.
28. C. Ajmi, S. E. Ferchichi, K. Laabidi, "New procedure for weld defect detection based-Gabor filter," in *2018 International Conference on Advanced Systems and Electric Technologies (IC_ASET)*, 2018, pp. 11–16.
29. O. Ronneberger, P. Fischer, T. Brox, "U-net: Convolutional networks for biomedical image segmentation," in *International Conference on Medical Image Computing and Computer-Assisted Intervention*, Springer, 2015, pp. 234–241.
30. W. Hou, Y. Wei, Y. Jin, C. Zhu, "Deep features based on a DCNN model for classifying imbalanced weld flaw types," *Measurement*, vol. 131, pp. 482–489, 2019.
31. D. Mery, V. Riffo, U. Zscherpel, G. Mondragón, I. Lillo, I. Zuccar, H. Lobel, M. Carrasco, "GDXray: The Database of X-ray Images for Nondestructive Testing," *Journal of Nondestructive Evaluation*, vol. 34, no. 4, pp. 1–12, 2015.
32. C. Ajmi, J. Zapata, S. Elferchichi, A. Zaafouri, K. Laabidi, "Deep Learning Technology for Weld Defects Classification Based on Transfer Learning and Activation Features," *Advances in Materials Science and Engineering*, vol. 2020, 2020.
33. A. Krizhevsky, I. Sutskever, G. E. Hinton, "Imagenet classification with deep convolutional neural networks," *Advances in Neural Information Processing Systems*, vol. 25, pp. 1097–1105, 2012.
34. R. Miao, Z. Jiang, Q. Zhou, Y. Wu, Y. Gao, J. Zhang, Z. Jiang, "Online inspection of narrow overlap weld quality using two-stage convolution neural network image recognition," *Machine Vision and Applications*, vol. 32, no. 1, pp. 1–14, 2021.
35. K. Simonyan, A. Zisserman, "Very Deep Convolutional Networks for Large-Scale Image Recognition," arXiv preprint arXiv:1409.1556, 2015.
36. K. He, X. Zhang, S. Ren, J. Sun, "Deep Residual Learning for Image Recognition," arXiv preprint arXiv:1512.03385, 2015.

37. Q. Wang, W. Jiao, P. Wang, Y. Zhang, "A tutorial on deep learning-based data analytics in manufacturing through a welding case study," *Journal of Manufacturing Processes*, vol. 63, pp. 2–13, 2021.
38. A. Paszke et al., "PyTorch: An Imperative Style, High-Performance Deep Learning Library," in *Advances in Neural Information Processing Systems 32*, H. Wallach et al., Eds. Curran Associates, Inc., 2019, pp. 8024–8035.
39. D. Mery, C. Arteta, "Automatic Defect Recognition in X-Ray Testing Using Computer Vision," in *2017 IEEE Winter Conference on Applications of Computer Vision (WACV)*, 2017, pp. 1026–1035.
40. S. J. Oh, M. J. Jung, C. Lim, S. C. Shin, "Automatic detection of welding defects using faster R-CNN," *Applied Sciences (Switzerland)*, vol. 10, no. 23, pp. 1–10, 2020.
41. R. Guo, H. Liu, G. Xie, Y. Zhang, "Weld Defect Detection from Imbalanced Radiographic Images Based on Contrast Enhancement Conditional Generative Adversarial Network and Transfer Learning," *IEEE Sensors Journal*, 2021.
42. L. Yang, H. Wang, B. Huo, F. Li, Y. Liu, "An Automatic Welding Defect Location Algorithm based on Deep Learning," *NDT & E International*, p. 102435, 2021.
43. M. Mirza, S. Osindero, "Conditional Generative Adversarial Nets," arXiv preprint arXiv:1411.1784, 2014.
44. T. Gantala, K. Balasubramaniam, "Automated Defect Recognition for Welds Using Simulation Assisted TFM Imaging with Artificial Intelligence," *Journal of Nondestructive Evaluation*, vol. 40, no. 1, pp. 1–24, 2021.
45. A. Bochkovskiy, C.-Y. Wang, H.-Y. M. Liao, "YOLOv4: Optimal Speed and Accuracy of Object Detection," arXiv preprint arXiv:2004.10934, 2020.
46. Y. Yan, D. Liu, B. Gao, G. Y. Tian, Z. C. Cai, "A Deep Learning-Based Ultrasonic Pattern Recognition Method for Inspecting Girth Weld Cracking of Gas Pipeline," *IEEE Sensors Journal*, vol. 20, no. 14, pp. 7997–8006, 2020.
47. J. Hu, W. Xu, B. Gao, G. Y. Tian, Y. Wang, Y. Wu, Y. Yin, J. Chen, "Pattern Deep Region Learning for Crack Detection in Thermography Diagnosis System," *Metals*, vol. 8, no. 8, 2018.
48. "API 5L : Specification for Line Pipe," American Petroleum Institute, Washington, D.C, U.S.A, 2004.
49. "API 1104 : Standard for Welding of Pipelines and Related Facilities," American Petroleum Institute, Washington, D.C, U.S.A, 2001.
50. G. Ghiasi, T.-Y. Lin, Q. V. Le, "Nas-fpn: Learning scalable feature pyramid architecture for object detection," in *Proceedings of the IEEE Conference on Computer Vision and Pattern Recognition*, 2019, pp. 7036–7045.
51. M. Tan, R. Pang, Q. V. Le, "Efficientdet: Scalable and efficient object detection," in *Proceedings of the IEEE/CVF Conference on Computer Vision and Pattern Recognition*, 2020, pp. 10781–10790.
52. M. Tan, Q. Le, "Efficientnet: Rethinking model scaling for convolutional neural networks," in *International Conference on Machine Learning*, PMLR, 2019, pp. 6105–6114.
53. J. Redmon, A. Farhadi, "Yolov3: An incremental improvement," arXiv preprint arXiv:1804.02767, 2018.
54. S. Naddaf-Sh, M-M. Naddaf-Sh, A. R. Kashani, H. Zargarzadeh, "An Efficient and Scalable Deep Learning Approach for Road Damage Detection," in *2020 IEEE International Conference on Big Data (Big Data)*, 2020, pp. 5602–5608.

55. T.-Y. Lin, P. Dollár, R. Girshick, K. He, B. Hariharan, S. Belongie, "Feature pyramid networks for object detection," in *Proceedings of the IEEE Conference on Computer Vision and Pattern Recognition*, 2017, pp. 2117–2125.
56. S. Ren, K. He, R. Girshick, J. Sun, "Faster r-cnn: Towards real-time object detection with region proposal networks," in *Advances in Neural Information Processing Systems*, 2015, pp. 91–99.
57. R. Girshick, "Fast r-cnn," in *Proceedings of the IEEE International Conference on Computer Vision*, 2015, pp. 1440–1448.
58. S. Liu, L. Qi, H. Qin, J. Shi, J. Jia, "Path aggregation network for instance segmentation," in *Proceedings of the IEEE Conference on Computer Vision and Pattern Recognition*, 2018, pp. 8759–8768.
59. J. Redmon, A. Farhadi, "YOLO9000: better, faster, stronger," in *Proceedings of the IEEE Conference on Computer Vision and Pattern Recognition*, 2017, pp. 7263–7271.
60. E. D. Cubuk, B. Zoph, D. Mane, V. Vasudevan, Q. V. Le, "Autoaugment: Learning augmentation strategies from data," in *Proceedings of the IEEE Conference on Computer Vision and Pattern Recognition*, 2019, pp. 113–123.
61. B. Zoph, E. D. Cubuk, G. Ghiasi, T.-Y. Lin, J. Shlens, Q. V. Le, "Learning data augmentation strategies for object detection," in *European Conference on Computer Vision*, Springer, 2020, pp. 566–583.
62. L. Perez, J. Wang, "The effectiveness of data augmentation in image classification using deep learning," arXiv preprint arXiv:1712.04621, 2017.
63. "COCO Detection Challenge (Bounding Box)," `https://cocodataset.org/#detection-eval`, Accessed: 2021-02-15.
64. T.-Y. Lin, P. Goyal, R. Girshick, K. He, P. Dollár, "Focal loss for dense object detection," in *Proceedings of the IEEE International Conference on Computer Vision*, 2017, pp. 2980–2988.
65. A. Howard et al., "Searching for MobileNetV3," arXiv preprint arXiv:1905.02244, 2019.
66. C.-Y. Wang, H.-Y. M. Liao, I.-H. Yeh, Y.-H. Wu, P.-Y. Chen, J.-W. Hsieh, "CSPNet: A New Backbone that can Enhance Learning Capability of CNN," arXiv preprint arXiv:1911.11929, 2019.
67. "EfficientDet (A PyTorch implementation of EfficientDet)," 2020, `https://github.com/rwightman/efficientdet-pytorch`.
68. I. Loshchilov, F. Hutter, "Sgdr: Stochastic gradient descent with warm restarts," arXiv preprint arXiv:1608.03983, 2016.
69. "Exponential Moving Average," `https://www.tensorflow.org/api_docs/python/tf/train/ExponentialMovingAverage`, Accessed: 2021-04-20.
70. "Nvidia Apex optimizers," `https://nvidia.github.io/apex/optimizers.html`, Accessed: 2021-01-30.
71. D. Hendrycks, K. Gimpel, "Gaussian Error Linear Units (GELUs)," arXiv preprint arXiv:1606.08415, 2020.
72. P. Ramachandran, B. Zoph, Q. V. Le, "Searching for activation functions," arXiv preprint arXiv:1710.05941, 2017.
73. D. Misra, "Mish: A self regularized non-monotonic neural activation function," arXiv preprint arXiv:1908.08681, 2019.
74. R. Wightman, "PyTorch Image Models," GitHub, 2019.

75. R. Padilla, S. L. Netto, E. A. B. da Silva, "A Survey on Performance Metrics for Object-Detection Algorithms," in *2020 International Conference on Systems, Signals and Image Processing (IWSSIP)*, 2020, pp. 237–242.
76. J. Redmon, S. Divvala, R. Girshick, A. Farhadi, "You Only Look Once: Unified, Real-Time Object Detection," arXiv preprint arXiv:1506.02640, 2016.
77. B. Zhou, A. Khosla, A. Lapedriza, A. Oliva, A. Torralba, "Learning deep features for discriminative localization," in *Proceedings of the IEEE Conference on Computer Vision and Pattern Recognition*, 2016, pp. 2921–2929.
78. R. R. Selvaraju, M. Cogswell, A. Das, R. Vedantam, D. Parikh, D. Batra, "Grad-cam: Visual explanations from deep networks via gradient-based localization," in *Proceedings of the IEEE International Conference on Computer Vision*, 2017, pp. 618–626.
79. Y. Wu, A. Kirillov, F. Massa, W.-Y. Lo, R. Girshick, "Detectron2," 2019, `https://github.com/facebookresearch/detectron2`.
80. G. Jocher et al., "ultralytics/yolov5: v5.0 - YOLOv5-P6 1280 models, AWS, Supervise.ly and YouTube integrations," Zenodo, Apr. 2021.

Section III

Arc Stud Welding

7 Explainable Defect Classification in Arc Stud Welding

Sadra Naddaf-Sh, M-Mahdi Naddaf-Sh, Maxim Dalton, Soodabeh Ramezani, Amir R Kashani, Hassan Zargarzadeh

CHAPTER SUMMARY

Arc Stud Welding (ASW) is widely used in many industries such as automotive and shipbuilding and is employed in building and jointing large-scale structures. While defective or imperfect welds rarely occur in production, even a single low-quality stud weld is the reason for scrapping the entire structure, financial loss, and wasting time. Preventive machine learning-based solutions can be leveraged to minimize the loss. However, these approaches only provide predictions rather than demonstrating insights for characterizing defects and root cause analysis. In this work, an investigation of defect detection and classification to diagnose the possible leading causes of low-quality defects is proposed. Moreover, an explainable model to describe network predictions is explored. Initially, a dataset of multi-variate time series of ASW utilizing measurement sensors in an experimental environment is generated. Next, a set of techniques to leverage synthetic measurements, reference, and residual signals, and generate a residue dataset, are proposed. Finally, the architecture of classification models is optimized and by the Bayesian black-box optimization methods to maximize their performance. Our best approach reaches an F1 score of 0.84 on the test set. Furthermore, an explainable model is employed to provide interpretations on per class feature attention of the model to extract sensor measurement contribution in detecting defects as well as its time attention.

7.1 INTRODUCTION

Arc Stud Welding (ASW) is a standard process in many industrial production lines, such as the automobile industry, and shipbuilding[1], where a metal fastener called Stud is welded to a workpiece with an electric arc [2]. The process includes heat transfer, mass transfer, metallurgical reaction, element diffusion, micro-structure change, and variation of mechanical properties, all of which make arc stud welding a complicated process [3]. Additionally, the process is characterized by high currents and short weld times. As the process is used in high-volume manufacturing production lines and is applied for structures with numerous stud welded joints, defective

DOI: 10.1201/9781003766575-7

or low-quality welds are inevitable. Therefore, a single defective weld leads to the rejection of an entire large-scale structure which imposes significant extra cost and time on the manufacturing process in high volumes [2, 3]. Hence, root cause analysis of low-quality or defective welds can result in minimizing disposed structures, financial loss, and optimization of the production line. Similar approaches [5] can be further applied to optimize costs.

Many factors result in a low-quality or defective weld stud, including weld area surface condition or material, inappropriate weld settings, malfunctioning equipment, and inexperienced welding operators [6]. While many are being avoided by appropriate supervision of the process, predicting a possible defective or substandard stud leads to saving time and costs. Furthermore, detecting precise causes helps experts prevent further issues. While in post-weld Non-destructive testing (NDT) X-ray imaging and automated analysis are applicable [7, 8], such techniques are not practical for complicated structures in automotive where uncounted weld studs appear in hard-to-access areas. Hence, the monitoring of the process and preventive techniques are more applicable. However, such data and the design of the experiment are not publicly available, and most welding setups are only equipped with voltage and current sensors, where no additional sensors are leveraged for monitoring.

In this study, a set of experiments are designed to synthetically generate defective arc stud welds while various sensors monitor and record welding parameters to generate a dataset of multivariate time-series (MTS) measurements for ASW. To enhance the detection process, a set of reference signals are logged, and synthetic measurements are generated with residual signals to shape a residue dataset. Finally, enhancements of the framework are empirically shown, and a comparative analysis of the methods for the classification of defects is investigated and presented.

Contributions of this work are as follows:

- An extended dataset of experiments is generated based on four sensor measurements during ASW for the purpose of monitoring the operation.
- A new algorithm to calculate the mean approximation of time-series signals and signal residuals is proposed, leading to the enhancement of the overall processing and classification model performances.
- A set of synthetic time-series measurements (e.g., Power) are generated and leveraged to provide enhanced measurements for models and to improve defect (root cause) classification results.
- A set of state-of-the-art classifiers are investigated and optimized for both their architectures and hyper-parameters employing the Bayesian methods in order to maximize the performance for the mentioned task. In addition, a comparative analysis of their performance is reported.
- Performance of an explainable deep MTS classification model is investigated and optimized to provide interpretation on how the model decides on features to classify signals as well as its time attention.

The remainder of this chapter is organized as follows. Section 7.2 described related works in this scope. Section 7.3 gives an overview of the terms in the chapter, the applied method for synthetic measurement and residual generation steps as well as network architectures. In section 7.4, the data preparation and dataset, training, evaluation, as well as experimental results, are explained in detail. Finally, in section 7.5 a discussion on performance is proposed, and in section 7.6, the chapter is concluded, and future improvements are presented.

7.2 RELATED WORK

In the past decades, few works are published for the enhancement of the ASW process with AI-powered technologies. In [1], an offline evaluation of parameter distribution during ASW is performed for seven trials with the purpose of common defect detection. Several experiments with various welding conditions, such as a surface with primer or rust, were designed, and parameters such as voltage, current, power, and resistance were measured. Results based on statistical analysis illustrate that, in most abnormal welding conditions, increased variation for voltage and current for fewer cases, e.g., surface with primer, is observed. The study is conducted by manual analysis of measurements.

In [2], quality monitoring and stability of the ASW process are investigated. Moreover, methods for real-time NDT for defect prediction are presented. Researchers present two neural net designs for monitoring weld quality. Welding Time, Welding Current Range, Workpiece Thickness, and Stud Diameter are used as inputs for both networks. One network has two outputs of Welding Current Peak Value and Torque at Failure, and the second network has a single output of visual Inspection (either defective or not). For both networks, a similar training set with only twenty training data was used. In the end, the relationship between the current peak and welding condition was observed and proved the advantage of utilizing the current peak for monitoring purposes. Moreover, employing these networks show the reading of additional weld quality-related parameters. Finally, all the approaches are tested on a tiny dataset with less than twenty samples, and none of them led to imperfection classification. In another work, ASW joint connection status is evaluated by ultrasonic technique [9], where A-scan signals are manually analyzed through Wavelet Packet. The authors concluded A-scan characteristics that relate to weak connection zones in stud welds.

More recently, researchers focused on addressing the imbalance issue due to the significant rare occurrence of anomaly cases for the task of fault prediction. In [10], the authors focus on proposing a new loss function for a recurrent neural architecture that handles imbalance and is specialized for multi-label fault prediction. In another work [11], Generative Adversarial Networks (GANs) are employed to learn patterns of anomaly cases in the frequency domain. Later, it is shown that Wasserstein GANs show significantly lower reconstruction error in comparison with variational Auto Encoder.

For supervised time series classification (TSC) and detection of MTS, many studies are done. In [12], Wang et al. presented a baseline for TS classification. Three baselines are presented and compared with older existing ones through 44 benchmark datasets. The baselines are Fully Convolutional Network (FCN), Residual network (ResNet), and deep multilayer perceptron (MLP). In most cases, these architectures outperformed older methods such as Dynamic time warping (DTW) [13], The bag-of-features framework (TSBF) [14], and Bag-of-SFA-Symbols (BOSS) [15]. In [16], a universal encoder is proposed that reaches state-of-the-art performance on MTS benchmark datasets. Finally, In [17] a comprehensive review on the classification of both univariate and MTS signals is done by applying both mentioned architectures and a few other designs existing in the literature with competing results on benchmark datasets. The work illustrates that ResNet, FCN [12], and Encoder [16] outperform other designs for MTS datasets. These models are appropriate choices for the baseline of our work.

Although many of the available classification methods in the literature have reached significant accuracy on benchmark datasets, a lack of accurate model explainability is discernible. In [12, 17] class activation maps (CAM) for univariate datasets are provided. However, in view of architectural limitations, unique designs are required to provide extended explanations for model behavior. In [18], an end-to-end explainable CNN design (MTEX-CNN) is provided to employ gradient-based methods for extracting the time and feature attention of the network on MTS datasets. Fauvel et al. redesigned the network for more accurate performance in prediction results and also significantly has fewer parameters (around ten times less) while boosting the accuracy of explanations of the model called XCM [19]. In this work, five models of MLP, FCN, ResNet, Encoder, and XCM are optimized and evaluated for ASW defect classification.

7.3 FRAMEWORK

In an industrial production line, because of the complexity and temporal compression of the process, gathering labeled data on defective welds is impractical. In fact, a non-destructive examination of the entire welded structure is performed afterward. Hence, the design of the experiment is required to carefully simulate similar conditions to reproduce common defective stud welds and generate a dataset of measurements.

In this work, convolutional classification architectures with novel residual signal and reference and synthetic measurement generation of multivariate time-series sensor measurements are utilized, investigated, and optimized. First, signal residuals are calculated based on either an existing reference signal or mean reconstruction of signals using Independent Component Analysis (ICA) whenever a reference signal does not exist. Second, residuals are aligned with original measurements. Moreover, the power and resistance of the process are added by multiplying and dividing the voltages by the corresponding currents, [1] followed by dynamic time-warping alignment of the resulting signals. Third, Bayesian optimized neural/convolutional architectures are used to perform classification. Finally, network output determines the root causes found in the weld MTS data. The entire system is presented in Figure 7.1.

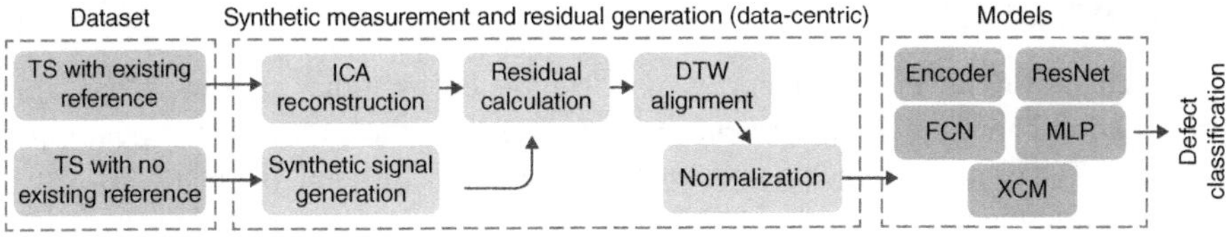

Figure 7.1 Block Diagram of Defect (root cause) Classification for Arc Stud Welding Process.

In the following, definitions used in the chapter are described, then mentioned steps are elaborated in detail.

7.3.1 DEFINITIONS

The technical terms used in this chapter are as follows:

- **Time-series**: Consider the time-series of T as a temporal ordered set of n variables sampled with a specific frequency of f and indexed with i to the last value, in the following equation [17]:

$$T = t[1], t[2], ..., t[n] \tag{7.1}$$

which is called univariate time series. On the other hand, an M-dimensional univariate Time-series collection is called Multi-variate Time-series (MTS) and is shown as:

$$T = T[1], ..., T[M] \tag{7.2}$$

Where each T is a univariate time series and M is equal to the number of dimensions [17].

- **residuals**: A residual of time-series T is calculated by $T' = T - \hat{T}$ where $\hat{T}$ is the mean approximation of T or a reference time-series which accounts for the mean approximation of it.

7.3.2 NORMALIZATION

Each measurement during the process is a T sampled with a frequency of f, and all values per each measurement type and stud type get normalized to zero mean and unit of energy before passing to the next stage using [20]:

$$\bar{T}[i] = \frac{T[i] - \mu}{\sigma}, \tag{7.3}$$

where μ is the mean of all measurements with the same type of measurement (e.g., all voltage measurements) and similar stud type, and σ is the standard deviation of the values.

7.3.3 RESIDUALS

For each univariate measurement, T gets subtracted by either its own mean approximation $\hat{T}$ or an exiting reference measurement for the very measurement type $T' = T - \hat{T}$. For measurements with an existing reference signal, the residual calculation is a simple one-by-one subtraction of features for each instance. However, for other signals, a $\hat{T}$ is required. Both ICA reconstruction and signal average were experimentally tested as a mean approximation. However, the mean approximation outperformed the average method in experiments. Next, the mathematical background of ICA is discussed.

7.3.4 RECONSTRUCTION WITH INDEPENDENT COMPONENT ANALYSIS

In this part, first ICA and fast ICA are briefly introduced, and then, the procedure of using ICA is described.

7.3.4.1 ICA

Independent component analysis (ICA) can separate or recover unknown independent sources from signal mixture observation [21, 22]. Let T be t[1], ..., t[n] observations with centering and whitening as a pre-processing step. Then, ICA can be written as

$$T = AS, \tag{7.4}$$

where A is the transformed mixing matrix [22] and S is a matrix that contains k independent components of T, where $k \leq n$ and each component is assumed to be non-Gaussian. ICA tries approximate A by maximizing the Independence of S, and the background assumption is that components of S are independent statistically. After acquiring A, the inverse matrix of it W is calculable, and then

$$S = WT. \tag{7.5}$$

As mentioned, a basic assumption for the ICA is non-Gaussianity. A classical method to measure non-normality is excess kurtosis which is computationally efficient and theoretically straightforward. Nonetheless, kurtosis is sensitive to outliers which makes it non-robust for non-Gaussianity measurements [22]. Instead, in the literature, the Negentropy method is used which is based on information theory entropy [22].

Entropy explains the degree of freedom of information for a given variable. Considering Y as a discrete random variable and, p as probability, then the entropy of H is:

$$H(Y) = -\sum p(Y) \log p(Y). \tag{7.6}$$

J, the Negentropy, is to obtain non-normality measurement, and it is similar to differential entropy value of which is always non-negative, and for Gaussian variables (i.e., Y_{gauss} of the same Y matrix) is equal to zero. J can be shown as

$$J(Y) = H(Y_{gauss}) - H(Y) \tag{7.7}$$

In [22] an efficient approximation for Negentropy is proposed, which is used in the following FastICA and is written as:

$$J(Y) \propto \{E[G(Y)] - E[G(v)]\}^2, v \sim N(0,1), \tag{7.8}$$

where $G(x) = \log\cosh(x)$ that is non-linear and non-quadratic.

7.3.4.2 FastICA

The efficient Negentropy J used to estimate W in (7.8) is called FastICA. With the assumption of having T centered and whitened, FastICA can estimate W that in [22], it is shown W estimations are consistent.

Finally, as in (7.5) an approximation of unknown sources S is generated, and by remixing with A in (7.4), an approximation of original time-series signals is achieved, which is used as a reconstruction of original signals. Practical employment is described more in-depth in the section 7.4.

7.3.5 DYNAMIC TIME WARPING

Dynamic Time Warping (DTW) is largely used in MTS, such as speech signals as well as classification tasks [23]. In this work, DTW is used to find an optimal global alignment of query and reference signals. It uses a Dynamic Programming (DP) matching approach to generate a distance matrix and pair-wise matching of each sample of the query TS and the nearest sample from the reference signal [24]. For this task, euclidean distance and symmetric step pattern with slope constraint condition of $P = 0$ are applied. DP equation (i.e., $g(i,j)$) for initial point is:

$$g(1,1) = 2d(1,1). \tag{7.9}$$

and for the rest is:

$$g(i,j) = \min \begin{bmatrix} g(i,j-1) + d(i,j) \\ g(i-1,j-1) + 2d(i,j) \\ g(i-1,j) + d(i,j) \end{bmatrix}, \tag{7.10}$$

where d is the distance function (i.e., $d(i,j) = \|a_i - b_i\|$).

7.3.6 NETWORK ARCHITECTURES

As mentioned in7.2, five models are applied and optimized in this work, as shown in Figure 7.2.

- **MLP**: Multi-layer Perceptron consists of stacks of fully connected dense layers followed by an activation layer to increase non-linearity and prevent saturation of the gradients. Also, a dropout layer is used at the layer's input to prevent overfitting, and the model ends with a softmax layer [12]. In our experiments, the number of layers, activation function, number of nodes in each dense layer, and dropout rate are considered hyperparameters and get optimized for MLP.

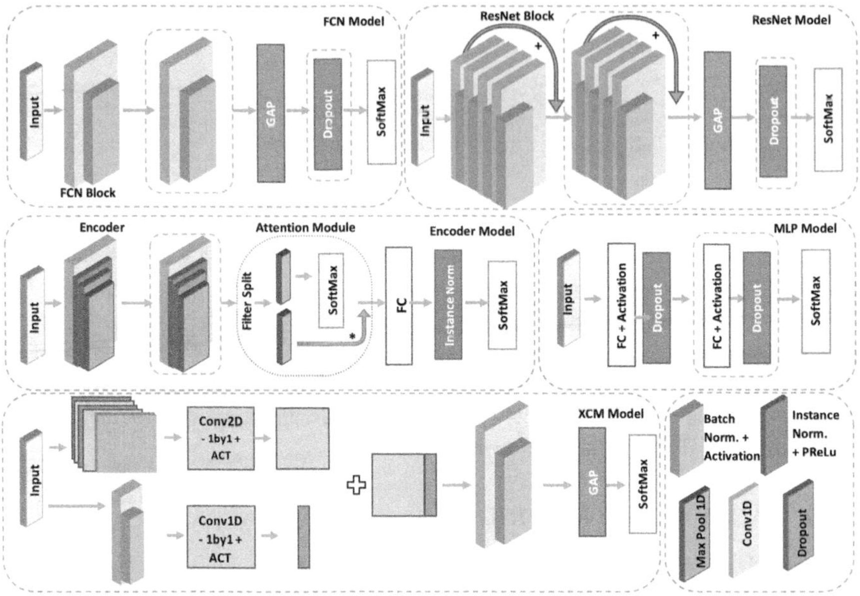

Figure 7.2 The models' architecture. The blocks inside Lime Dashed lines are repeated blocks, and the number of repeats can from 1 to 3. The optimal repeat number is searched for during optimization.

- **FCN**: FCN consists of a few repeated blocks followed by a Global Average Pooling (GAP) layer instead of a fully connected layer, and has a softmax layer as the final layer. Each block starts with a 1-Dimension Convolutional layer followed by a batch normalization layer which speeds up the training process and reduces over-fitting, and an activation layer [12]. In our optimization pipeline, the number of blocks, filter length, the number of filters in 1-D Convolution, and the type of activation function get optimized for this design. In addition, a few optional fully connected layers are added just before the softmax layer, and the number of these layers (can be zero), number of nodes, and dropout rate are to get optimized.
- **ResNet**: ResNet consists of a few residual blocks, and, similar to FCN, ends with GAP and softmax layers. Each residual block has the structure of a 1-D convolutional layer, Batch normalization, and activation layer three times. Also, lateral connections are added from the input of the block to the end of the same block [12, 26]. The number of blocks, filter length, the number of filters in 1-D Convolution, and the type of activation function get optimized for this design. Furthermore, similar to FCN, optional dense layers are added, and their parameters get optimized during architecture optimization.

- **Encoder**: This design is inspired by FCN. However, the GAP layer is replaced with an attention layer [29] in which the network learns which parts of the time series are more important for classification. The encoder model also consists of a few repeated blocks; each block starts with 1-D convolution followed by instance normalization [28], Parametric ReLU [27] activation, and finally, a dropout. Each block is followed by max-pooling except the last layer, which gets fed into the attention layer, followed by a dense layer and a softmax. For this design, the number of repeated convolutional blocks, number of filters and filter length of 1-D convolutions and drop rate of dropout layer, activation of dense layer, and number of its nodes.
- **XCM**: This design starts with two separate paths. One path uses 2-D convolutions, batch normalization, and activation, and finally uses 1x1 convolution and activation. The second path is similar to path one, except it uses 1-D convolutions. Next, the results of these two paths get concatenated and then pass through one more 1-D convolution followed by GAP and softmax layer. Having two separate paths help to extract feature and time attention of the input using gradient-based methods. For this design, filter length, the number of filters, and activation function type gets optimized.

7.4 IMPLEMENTATION

Arc Stud Welding is described in [1, 2, 3, 4], and the same procedure (shown in Figure 7.3) is performed for experiments in this work, except the process is automated through a robotic arm. During the welding process, sensors in the robotic arm actuator monitor voltage and current. Simultaneously, the lift position sensor also monitors fastener positioning and the reference signal for lift position. LMcurrent is the current of the linear motor that moves the stud and is a critical factor in the penetration of the surface material. For these experiments: The average penetration is 0.70 with an average welding time of 72 ms. The sampling rate is approximately 1 kHz, and finally, all logged sensors' data get transmitted through MQTT protocol to generate a dataset of defective welds.

In the following, the dataset is described, and synthetic signal generations, as well as prediction models, are elaborated on.

7.4.1 DATASET

The initial dataset contains six classes of Arc Stud Welds with various welding conditions, as mentioned, and measurement types include Voltage, Current, LM Current, and its reference, lift position, and its reference (residual of which is called penetration). Additionally, synthetic measurements, including Power (i.e., pair-wise V*I for each feature) and Resistance (i.e., V/I), are also computed and added to the dataset. Therefore, there are six measurement types in total. Although Current is a controlling parameter of the process, experiments show that it helps network classify more accurately. Hence, it is kept among measurements.

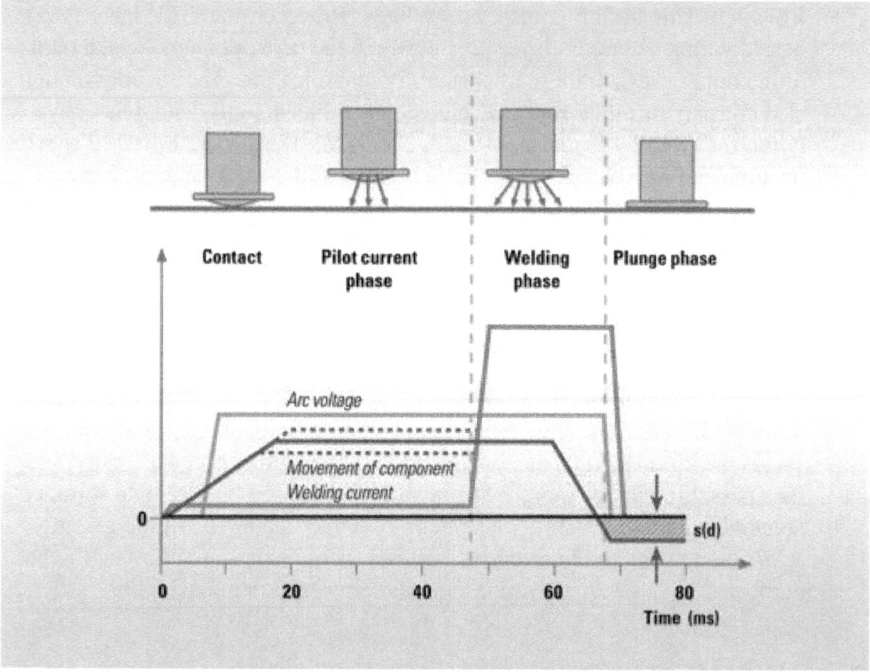

Figure 7.3 Arc Stud Welding Process [25].

As mentioned in this process, defective welds are synthetically generated with known root causes of the defects. The six types of defects (or root causes) are synthetically generated based on the design of the experiment principles. The defect types include four types of pollution on the welding surface which are coded into Root Cause (RC) 1 to Root Cause (RC) 4. Two types of defects are also made on stud quality and electrical system and coded into Root Cause 5 and Root Cause 6. Figure 7.4 visualizes how measurements for various Root Causes across classes are similar, illustrating how patterns are complicated to extract, and why additional synthetically generated measurements and residual signals are helpful. The final dataset has 263 samples from 6 classes, and the distribution is shown in Table 7.1. Each instance has six measurements with 480 timestamps. The final dataset (i.e., residue dataset) has six measurements of Voltage residual, Current residual, Penetration, LMCurrent residual, Power, and Resistance.

7.4.2 PRE-PROCESSING

Initially, only voltage and current have a length of 480. Other measurements are padded with zeros to have the same length. For LM current, at the end of the

Table 7.1
Class Distribution for Dataset.

Classes	*RC1*	*RC2*	*RC3*	*RC4*	*RC5*	*RC6*
Count	59	60	56	52	16	20

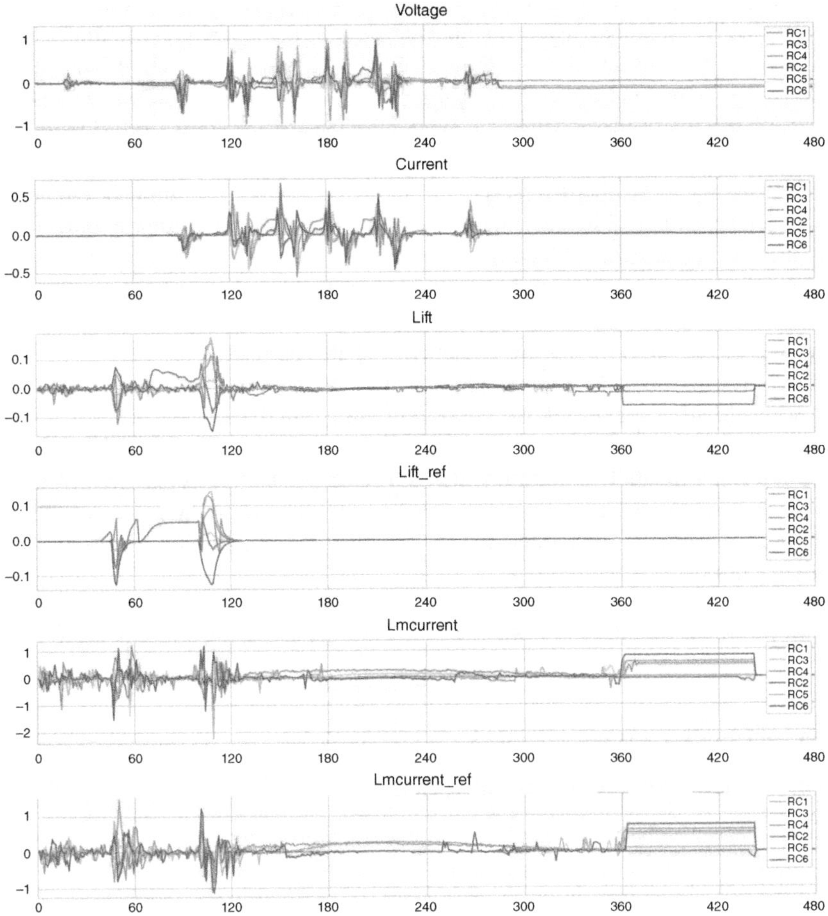

Figure 7.4 The similarity of measurements across classes RC1 to RC6. The full length of signals is visualized (i.e., 0 to 480).

process, it is stopped, and thus zero-padding is meaningful. For the lift position, since the nozzle turns back to the initial point, then padding with zero is rational and meaningful.

7.4.3 SYNTHETIC MEASUREMENT AND RESIDUAL GENERATION

As described in 7.3, a series of steps are performed to generate synthetic residuals to use as a reference for voltage and current. Starting with a mean approximation of signals to get residuals. Since for LM and lift position a reference signal exists, the existing reference will be used as $\hat{T}$ and residual (T') is $T' = T - \hat{T}$. For measurements with no existing reference signal, a mean approximation is required. As mentioned in 7.3.4, since independent components of unknown sources of S in (7.5) are an approximation of the original signal T, remixing through (7.4) creates an approximate of the original signal with common unknown sources. Hence, the resulting signal $(\hat{T})$ is employed for creating a residual signal. Therefore, to acquire mean reconstruction ICA, for measurements with no reference signal (i.e., voltage, power, resistance) and per stud type, a three-component ICA transform and inverse transform (reconstruction) are done, respectively. Then, the mean of reconstructed signals (all signals have the same stud type, measurement type, and features) is calculated and used as $\hat{T}$ to get T' residual. For synthetic data (i.e., power and resistance), multiplication or division is done prior to residual calculation. Finally, FastICA from [30] implementation is applied for the mentioned reconstructions.

Next, dynamic time warping is applied to the signals. Since the start and the end of extreme periods or peaks change along with signals, then extreme periods in the query signal and mean signal do not overlap, and this results in shifts in the final residual. Hence, using DTW on residual (T') as the original signal and (T) as a reference would align peaks with the query reference signal. Note that for signals with existing reference signals, DTW is not required, for no shifts appear in those. Moreover, it is experimentally determined that applying symmetric step pattern results in a smaller rooted mean square error in comparison with an asymmetric step pattern. Finally, Algorithm 1 summarizes how the residual calculation is done to create the residue dataset.

As the last step, normalization is done per sensor measurement type using (7.3). Figure 7.5 shows samples of normalized voltage and current, its mean reconstruction, and aligned and non-aligned residual signals.

7.4.4 CLASSIFICATION MODELS

7.4.4.1 Hyper-parameter Optimization

Residue dataset gets fed into several architectures described in section 7.3.6, and shown in 7.2 including MLP, FCN, ResNet, Encoder, and XCM. The models have a pre-defined number of blocks, and layers as described in the original works. To show the effectiveness of Algorithm 1, the models are once trained with default parameters as used in [17], and then systematically optimized. For optimization, all hyper-parameters and overall architectures are optimized using the Bayesian Methods. Table 7.2, illustrates search space for the Bayesian Hyperparameter Optimization (HPO). Max-pooling is only used for the Encoder model, while the rest of the parameters are common for all models.

Algorithm 1: residual calculation for a specific sensor measurement type

Result: residuals of T' as residuals_set ;
residuals_set ← [ ] ;
if *reference signal* ($\hat{T}$) *exists* **then**
  while *not processed every instance* T **do**
    $T' \leftarrow T - \hat{T}$;
    residuals_set ← T' ;
  end
else
  ICA,$\hat{T}_{mica}$ ← [] ;
  ICA ← fit ICA for the entire set of Ts ;
  ICA ← Inverse Transform ICA ;
  $\hat{T}_{mica}$ ← calculate mean of reconstructed signals over x-axis;
  while *not processed every instance* T **do**
    $T \leftarrow DTW(T, \hat{T}_{mica})$;
    $T' \leftarrow T - \hat{T}_{mica}$;
    residuals_set ← T' ;
  end
end
normalize residuals_set using (7.3);

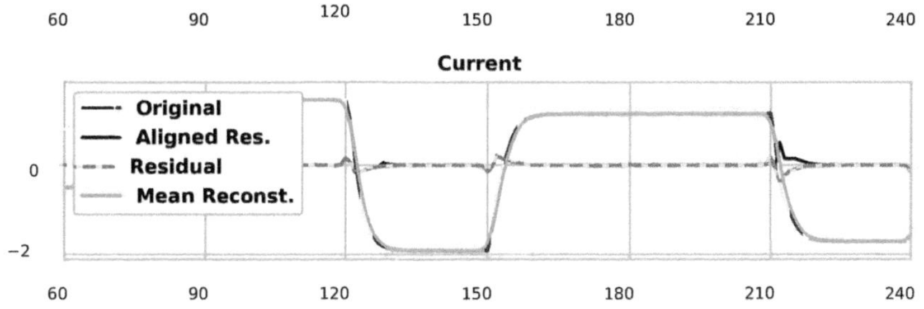

Figure 7.5 Samples of voltage and current for original normalized measurements, their ICA mean Reconstructions for a specific sample, their residuals, and DTW aligned residuals. Only timestamps 60 to 240 are visualized for a clear demonstration.

Table 7.2
Hyper-parameters search domain used for Bayesian Optimization.

Parameters	Search Space
optimizer	Adam, SGD, RMSprop, Adadelta
activation functions	relu, swish, sigmoid, tanh, selu
# of layers/blocks	2 to 5
# of filters	16, 32, 64, 128, 256
kernel size	1 to 12 (dependent to model)
drop out rate	0 to 0.5 with 0.1 step
Batch size	32 to 256 with 32 step
Epochs	40 to 100
learning rates	1e-2 to 10e-4 (logarithmic)
max-pooling size	2 & 4 (for Encoder only)
Dense layer nodes	128 & 512 (when applicable)

7.4.4.2 Training Setting

All models are optimized and tested on a single NVIDIA GTX 1080 GPU. The implementation and HPO are done using Tensorflow v2.3 and Keras v2.1. Keras-Tuner [31] is used for HPO. Keras-Tuner is a Keras-friendly framework, with implementations of HPO algorithms. It is possible to smoothly manage to optimize activation functions, kernels, layers, and architecture of a descriptive Keras implementation. For the experiments, Bayesian optimization is used with the Gaussian Process model as the surrogate model, and the acquisition function used is the upper confidence bound (UCB). Also, the objective was set to maximize validation set accuracy, with 50 initial points.

Single model training is performed in two to three minutes; However, HPO takes between two to five hours depending on model size, number of parameters, and search space. Finally, inference time is around a millisecond (ms) for a batch size of one and two ms per batch with a size of 32, meaning the model is able to predict in real time.

7.4.4.3 Optimized Models

The residue dataset contains six measurements, which are padded to have similar lengths. Thus, the models are fed with tuples with shape (batch size (bs), sequence length, number of signals), i.e. (bs, 480, 6). 80% of the samples of the dataset stratifiedly sampled for training and 5-fold cross-validation, and the remaining 20% is withheld for unbiased testing. Early Stopping with the patience of 15 epochs is also applied. For models trained with the raw dataset, the learning rate is fixed to 1e-3 with Adam optimizer, batch size of 32, and relu as activation function, the rest of the

Table 7.3
Hyperparameters extracted for various architectures were extracted using Bayesian Optimization.

Parameters	Models				
	MLP	FCN	ResNet	Encoder	XCM
Optimizer	Adam	Adam	Adam	Adam	Adam
activation fun	tanh	relu	swish	relu	relu
learning rate	10e-4	10e-2	10e-4	10e-4	10e-4
Epochs	100	100	100	100	100
Batch size	32	32	32	32	32
# of blocks	4	3	3	2	2
# maxpooling	-[a]	-	-	2	-
# of filters	-	16|256	32|64|64	256|512	128|128
kernel sizes	-	12|12	6|4|9	5|12	20|20
Dropout	0.4|0.5|0.1	0.2|0.5	-	0.5	-
Dense	350|350|500	128|32	-	128	-
# of param.	1.3M	89K	257K	1.7M	38K

[a] denotes that the parameter does not exist for the specified model.

parameters can be found in [17]. Table 7.3 summarizes the best case models for each design.

Table 7.4 reveals average performance, recall, F1-score for all four models.

Table 7.5 illustrates the performance of the Encoder model as the best model for each class. As it shows, the hardest root cause to detect is RC4.

7.4.4.4 Results Explainability

As described, XCM is designed to be capable of providing explainability for multivariate time-series datasets. Figure 7.6 shows a sample of saliency map for a True positive prediction from class RC2, in which the red areas show high attention to the network, while white and blue show no to little attention. Figure 7.6-a describes that most of the attention of the model is to the detection of a defect from RC2 is related to the mid-part and ending part of penetration. Nonetheless, the network is not considering current and voltage residual (i.e., current* and voltage*), and the minimum consideration is for Resistance and Power. Figure 7.6-b relates to time attention, and it is related to the combinatorial path of the network (i.e., the path with 1-D convolutions), which represents that mid-part of the signal is taken into consideration for RC2 defect detection and relates to the second cycle of the welding process.

Table 7.4

Models performance on test/validation sets. F1-score is weighted to address class imbalance on final accuracy.

	Models	Test Set			Validation Set
		Precision	**Recall**	**F1-score**	**F1-score[a]**
raw dataset	MLP	0.817	0.807	0.805	0.76
	FCN	0.66	0.596	0.551	0.516
	ResNet	0.59	0.654	0.616	0.52
	Encoder	0.834	0.769	0.767	0.759
	XCM	0.65	0.53	0.52	0.51
residue dataset (our method)	MLP	0.81	0.807	0.799	0.853
	FCN	0.682	0.71	0.668	0.776
	ResNet	0.73	0.71	0.71	0.75
	Encoder	0.847	0.847	**0.843**	0.849
	XCM	0.59	0.57	0.58	0.56

[a] These values are averaged F1-score of validation over 5-fold.

Table 7.5

Metrics per class for best model (Encoder).

Class	Test Set		
	Precision	**Recall**	**F1-score**
RC1	0.85	1.0	0.92
RC2	0.70	0.70	0.70
RC3	0.73	0.67	0.70
RC4	1.0	0.67	0.80
RC5	1.0	1.0	1.0
RC6	1.0	1.0	1.0
Weighted Average	**0.85**	**0.85**	**0.84**

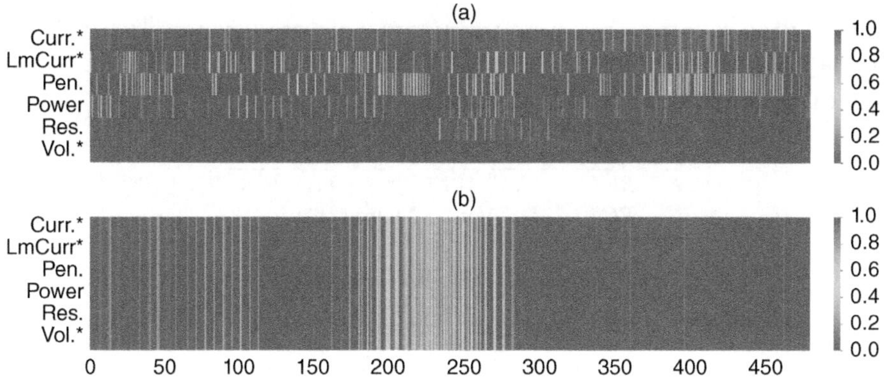

Figure 7.6 Feature Attention-a, Time Attention-b for prediction of class RC2. the x-axis is timestamps and the y-axis is features, Current*, LmCurrent, Penetration, Power, Resistance, Voltage*. All are residual signals. Penetration is the difference between lift position and its reference, * underlines that residual is used.

7.5 DISCUSSION

7.5.1 MODEL PERFORMANCE

Table 7.3 shows the best parameters extracted using the Bayesian method. As demonstrated for all models, Adam is the best optimizer, and in most cases, ReLU showed high performance as an activation function. As mentioned in Figure 7.6, optional fully connected layers were added after the GAP layer and based on Table 7.3, these show noticeable improvement in overall network performance. Moreover, models FCN, Encoder, and ResNet tend to start with smaller window sizes and the number of filters, and these increase further on deeper layers.

Table 7.4 demonstrates model performance on both test sets and averaged F1-score of 5-fold on the validation set. It reveals that models are showing great performance on about 65% of training data, and the best model is the Encoder model. Table 7.5 shows per-class accuracy for all classes using Encoder Model. Classes related to RC5 and RC6 are the easiest to classify, while RC2, RC3, and RC4 are the hardest. Although the Encoder model shows a 0.70+ F1 score per class, the same does not happen with the per-class F1 score of other models. For instance, FCN shows a 0.85+ F1-score for classes RC2 and RC3, while it fails to detect RC4 class (i.e., less than 0.4 accuracy). Thus, a model ensemble will help to reach higher overall performance.

7.5.2 MODEL EXPLAINABILITY

XCM model feature attentions on the test set reveal feature importance in for each classification. Figure 7.6 demonstrates an example for RC2 class, which is a workpiece defect. The overall analysis of feature importance is summarized in Table 7.6, where features importance for each class is sorted in order of their contribution (i.e.,

Table 7.6
Per Class XCM Measurement Attentions.

Class	Measurement Attentions
RC1	Vol*, LMCurr*, Curr*, Res.
RC2	Vol*, LMcurr*, Curr*
RC3, RC4	-
RC5, RC6	Pen., Res.

left most feature is the one with the most contribution. Linear Motor current is the feature with the most contribution in detection. One explanation is, the impurity on the surface causes slight movements when the welding head makes contact with the surface, which results in variations in LMCurrent. These slight movements vary across classes, and are a reliable source for the model to make classification decisions. On the other hand, for stud-related impurities (i.e., RC5, RC6), the common discriminative factor is Penetration. Although the rest of the signals had partial contributions to classification, (illustrated in Figure 7.6), their existence is crucial to achieving the highest performance. A witness to this is the results in Table 7.4, where models without synthetic measurements have lower accuracy. Finally, since the XCM model performance is not acceptable for RC3, and RC4, they are not reported.

7.5.3 WELDING IMPROVEMENTS

The described procedure can appropriately and immediately detect issues related to the base material, stud, workpiece, or configuration, where slight degradation can cause unexpected issues. As the system and monitoring are standalone in the process, it does not have any interference with the welding operation. Moreover, this allows overtime monitoring of the system and components, where unexpected changes and downtime can be monitored and addressed during scheduled maintenance rather than causing unexpected emergency downtime, which increases productivity and prolongs the system's life span. Furthermore, adaptive weld parameter control is achievable based on surface impurity. For example, based on observed primer on the surface weld time should be optimized to increase or decrease energy transfer, and reduce surface errors.

7.6 CONCLUSIONS

In this chapter, an initial dataset of Stud Welding is generated and transformed into the residue dataset with a novel method to boost classification performance. Moreover, the Bayesian hyper-parameter optimization and comparative analysis of deep convolutional classifiers for MTS are conducted. Furthermore, the explainability of the model, feature, and time attention based on XCM architecture is analyzed.

Our experiments show the combination of our approach of synthetic measurements, residual signal generation, and deep classifier has enabled further investigations on tracking possible causes of stud weld failure. The residue dataset creation steps consist of the mean approximation of signal through ICA reconstruction, which is then used as a reference for calculating signal residual and is followed by performing dynamic time warping for signal alignment. For classifiers, MLP, FCN, ResNet, Encoder, and XCM are investigated and optimized both in terms of architecture and parameters. The most accurate model is the Encoder model, with an F1-score of 0.84 on both the 5-fold cross-validation and test set.

Although the approach reveals high-performance defect classification, further investigation is required for both improving existing designs and production deployment. For improving present designs, ensemble methods, extending the dataset, and investigation feature importance are suggested. Another interesting path can be a deep investigation of the explainability of models. XCM is chosen as it outperforms other existing models and utilizes ten times fewer parameters than MTEX-CNN. However, the explainability capabilities are still to be investigated. For production requirement, investigations for validating the same performance is suggested as production is not labeled, and hand labeling such data is either impossible in most cases or time-consuming and costly.

REFERENCES

1. Samardžić, I., Klarić, Š., & Siewert, T. (2007). Analysis of welding parameters distribution in stud arc welding. *Welding & Materials Technical, Economic and Ecological Aspects*, 791–800.
2. Al-Sahib, N. K. A., Ameer, H. K. A., & Ibrahim, S. G. F. (2009). Monitoring and quality control of stud welding. *Al-Khwarizmi Engineering Journal*, *5*(1), 53–70.
3. Hildebrand, J., & Soltanzadeh, H. (2014). A review on assessment of fatigue strength in welded studs. *International Journal of Steel Structures*, *14*(2), 421–438.
4. Ramasamy, S. R., Gould, J., & Workman, D. (2002). Design-of-experiments study to examine the effect of polarity on stud welding. *Welding Journal New York*, *81*(2), 19–S.
5. Heidarydashtarjandi, R., Prasad-Rao, J., & Groth, K. M. (2022). Optimal maintenance policy for corroded oil and gas pipelines using Markov decision processes. *International Journal of Prognostics and Health Management*, *13*(1).
6. Chambers, H. A. (2001). Principles and practices of stud welding. *PCI Journal*, *46*(5), 46–59.
7. Naddaf-Sh, S., Naddaf-Sh, M.-M., Zargarzadeh, H., Dalton, M., Ramezani, S., Elpers, G., Baburao, V. S., & Kashani, A. R. (2022). Real-time explainable multiclass object detection for quality assessment in 2-dimensional radiography images. *Complexity*, *2022*.
8. Naddaf-Sh, M.-M., Naddaf-Sh, S., Zargarzadeh, H., Zahiri, S. M., Dalton, M., Elpers, G., & Kashani, A. R. (2021). Defect detection and classification in welding using deep learning and digital radiography. In *Fault Diagnosis and Prognosis Techniques for Complex Engineering Systems* (pp. 327–352). Elsevier.
9. Dong, J., Xu, G., Yu, H., Fan, G., Wei, L., & Gu, X. (2019). Connection status evaluation in arc stud weld joints by ultrasonic detection. *The International Journal of Advanced Manufacturing Technology*, *100*(1), 663–672.

10. Zhang, W., Jha, D. K., Laftchiev, E., & Nikovski, D. (2019). Multi-label prediction in time series data using deep neural networks. *International Journal of Prognostics and Health Management*, *10*(4).
11. Ducoffe, M., Haloui, I., & Gupta, J. S. (2019). Anomaly detection on time series with Wasserstein GAN applied to PHM. *International Journal of Prognostics and Health Management*, *10*(4).
12. Wang, Z., Yan, W., & Oates, T. (2017). Time series classification from scratch with deep neural networks: A strong baseline. In *2017 International Joint Conference on Neural Networks (IJCNN)* (pp. 1578–1585). IEEE.
13. Keogh, E., & Ratanamahatana, C. A. (2005). Exact indexing of dynamic time warping. *Knowledge and Information Systems*, *7*(3), 358–386.
14. Baydogan, M. G., Runger, G., & Tuv, E. (2013). A bag-of-features framework to classify time series. *IEEE Transactions on Pattern Analysis and Machine Intelligence*, *35*(11), 2796–2802.
15. Schäfer, P. (2015). The BOSS is concerned with time series classification in the presence of noise. *Data Mining and Knowledge Discovery*, *29*(6), 1505–1530.
16. Serrà, J., Pascual, S., & Karatzoglou, A. (2018). Towards a universal neural network encoder for time series. In *CCIA* (pp. 120–129).
17. Fawaz, H. I., Forestier, G., Weber, J., Idoumghar, L., & Muller, P.-A. (2019). Deep learning for time series classification: A review. *Data Mining and Knowledge Discovery*, *33*(4), 917–963.
18. Assaf, R., Giurgiu, I., Bagehorn, F., & Schumann, A. (2019). MTEX-CNN: Multivariate time series explanations for predictions with convolutional neural networks. In *2019 IEEE International Conference on Data Mining (ICDM)* (pp. 952–957). IEEE.
19. Fauvel, K., Lin, T., Masson, V., Fromont, É., & Termier, A. (2020). XCM: An explainable convolutional neural network for multivariate time series classification. *arXiv preprint arXiv:2009.04796*.
20. Goldin, D. Q., & Kanellakis, P. C. (1995). On similarity queries for time-series data: Constraint specification and implementation. In *International Conference on Principles and Practice of Constraint Programming* (pp. 137–153). Springer.
21. Chen, K.-H., & Khashanah, K. (2015). The reconstruction of financial signals using fast ICA for systemic risk. In *2015 IEEE Symposium Series on Computational Intelligence* (pp. 885–889). IEEE.
22. Hyvärinen, A., & Oja, E. (2000). Independent component analysis: Algorithms and applications. *Neural Networks*, *13*(4-5), 411–430.
23. Górecki, T., & Łuczak, M. (2015). Multivariate time series classification with parametric derivative dynamic time warping. *Expert Systems with Applications*, *42*(5), 2305–2312.
24. Sakoe, H., & Chiba, S. (1978). Dynamic programming algorithm optimization for spoken word recognition. *IEEE Transactions on Acoustics, Speech, and Signal Processing*, *26*(1), 43–49.
25. Stanley Black and Decker. (2022). Stud welding products & systems. *STANLEY® Engineered Fastening*. Retrieved from http://www.emhart.eu/eu-en/products-services/products-by-category/tucker-stud-welding/stud-welding-systems/energy-units.php (Accessed: 04-26-22).
26. He, K., Zhang, X., Ren, S., & Sun, J. (2016). Deep residual learning for image recognition. In *Proceedings of the IEEE Conference on Computer Vision and Pattern Recognition* (pp. 770–778).

27. He, K., Zhang, X., Ren, S., & Sun, J. (2015). Delving deep into rectifiers: Surpassing human-level performance on ImageNet classification. In *Proceedings of the IEEE International Conference on Computer Vision* (pp. 1026–1034).
28. Ulyanov, D., Vedaldi, A., & Lempitsky, V. (2016). Instance normalization: The missing ingredient for fast stylization. *arXiv preprint arXiv:1607.08022*.
29. Bahdanau, D., Cho, K., & Bengio, Y. (2014). Neural machine translation by jointly learning to align and translate. *arXiv preprint arXiv:1409.0473*.
30. Pedregosa, F., Varoquaux, G., Gramfort, A., Michel, V., Thirion, B., Grisel, O., Blondel, M., Prettenhofer, P., Weiss, R., Dubourg, V., & others. (2011). Scikit-learn: Machine learning in Python. *The Journal of Machine Learning Research*, *12*, 2825–2830.
31. O'Malley, T., Bursztein, E., Long, J., Chollet, F., Jin, H., Invernizzi, L., & others. (2019). KerasTuner. Retrieved from https://github.com/keras-team/keras-tuner

8 Machine Learning in Automotive Stud Weld Defect Classification

Sadra Naddaf-Sh, M-Mahdi Naddaf-Sh, Maxim Dalton, Soodabeh Ramezani, Amir R Kashani, Hassan Zargarzadeh

CHAPTER SUMMARY

Arc stud welding is a well-developed and automated manufacturing process commonly used to fasten studs to a workpiece. This process requires implementation of measures to prevent low-quality welds that can lead to disposal of an entire structure. In this chapter, a machine learning-based method is introduced that uses experimental data to detect and classify defective stud welds. the source of data is a series of experimental arc stud welds conducted by a precision robotic arm. Welding data log is recorded through various sensors to shape a multivariate dataset for arc stud welding. Then, wavelet transforms as well as synthetic signal generation are applied for feature extraction. For each set of extracted features, suitable models are designed, trained, and evaluated for automatic root cause classification. Next, the models and approaches are compared on synthetic measurement generation, feature extractions, and feature importance. Numerical results show that XGBoost, the best model, has 85.41% F1-score on the test set. Additionally, the study reveals that voltage and lift play crucial roles in the root cause classification of arc stud weld. The proposed method not only offers promising results for defect detection in arc stud welding but also provides outreach value for similar industrial processes.

8.1 INTRODUCTION

The invention of arc stud welding (ASW) happened right before World War II to address the need for attachment of wood planking for aircraft carriers. This was an alternative to tedious, inaccurate, and labor intensive process of drilling holes. Since then, arc stud welding is the prevalent practice in construction and manufacturing processes such as automotive and shipbuilding, for joining metal studs to workpieces [1, 16, 3]. ASW process refinement is of improvements from two perspectives. First, the strength, quality, and integrity of joints across the structure in finished product determines the safety of product. Second, reducing costs, industrial waste is entangled with refinement and optimization of the welding practices. For instance, a single low-quality joint leads to rejection of whole structure, disposal, producing a great amount

DOI: 10.1201/9781003766575-8

of industrial waste, environmental damage, and financial loss. Although process automation has alleviated the faults with human error root causes, necessity for instant fault detection, and root cause analysis of faults is discernible.

Regardless of human errors, several parameters impact the quality of finished product and make the optimization process challenging. According to [3], weld area surface condition or material, stud weld material, weld settings, out-dated and malfunctioning equipment, and lack of quality control and inspection procedures, are the major root causes of faulty stud welds. Based on current procedures, non-destructive testing (NDT) is done to determine existence of faulty joints by subject matter experts. Since the procedure is a post-weld inspection, it does not have the capacity to determine the source root cause until after welding, which leads to propagation of faults. For example, due to slight variation in base material of the structure, e.g., a car body, defective weld starts from the very first joint, and propagates to the following joints, while it could be detected only after welding is done.

To eliminate laborious fault detection and enable automated manufacturing monitoring, various techniques have been introduced in the context of faulty join diagnostics detection and NDT inspection. This includes parameter distribution analysis [1], defective, non-defective classification of welded joints [2], spectral analysis of joints for online monitoring [23], and leveraging A-scans and wavelet packet analysis for determining weak connections at fusion zones [26].

A limited number of articles are published covering the ASW analysis, for the product is commercial, and economically critical to industries. In [1], seven trials on parameter distribution of ASW are conducted to perform defect detection. Each trial is done in different welding conditions, such as a surface with primer or rust. The results reveal that voltage has a key role in most abnormal conditions, while welding current measurement has a slight variation in the abnormal trials. Hence, the current is not a key factor in determining abnormal welded studs.

Authors in [2], investigated ASW defective, non-defective classification, stability, quality, and real-time NDT. For ASW stability, two fully connected artificial neural networks (ANNs) are designed, and both are fed and trained with welding time, welding current range, workpiece thickness, and stud diameter. One of the ANNs outputs welding current peak value and torque at failure, and the other has a binary output determining whether the weld is visually defective or non-defective. It was concluded that the current peak can be used for monitoring weld quality. In this study, models were trained with only 20 instances and feature selection is manual.

In [23], authors leverage spectral analysis and ANNs to provide online quality monitoring of arc weld seams by analyzing welding plasma spectra in real time. In [26], ASW joints connection quality is evaluated by ultrasonic technique, where A-scan signals are processed through wavelet packet. The authors reported that A-scan characteristics can be leveraged to distinguish between weak connections and fusion zones, which leads to the practice of NDT for ASW despite the manual nature of the analysis. Manual analysis is widely used, after feature selection for NDT. For example, in [30], eddy current pulsed thermography is employed for sampling and later multi-resolution wavelet transform combined with principal component analysis (PCA) on high-dimensional data, which resulted in smaller detect detection with visual observation.

A machine learning methodology to perform pattern recognition for time series measurements in a multivariate setting like ASW, is time series classification (TSC). Inspired by the emergence of novel techniques in deep learning, significant improvement is achieved for TSC task. In [8], Wang et al. presented three baselines of the fully convolutional network (FCN), residual network (ResNet), and deep multilayer perceptron (MLP). Fawaz et al. [4] reviewed both univariate and multivariate time series (MTS) classification on various architectures accomplishing competing results on benchmark datasets.

Although, architectures, attention mechanism, and training settings play crucial roles in the outcome of Deep networks, a parallel thread is investigation on providing high-quality features for models. In [24], 2D wavelet transforms are employed for damage detection in steady-state wave-field data of a laser scanning system. Authors experimentally determine the optimal mother wavelet based on corrosion on aluminum plates, and validation is done manually. Authors in [25], performed NDT on the content of composite materials by applying DWT. They used a five-level decomposition using Daubechies (db4) wavelet and performed four regression analyses resulting in a max rooted mean square error of 2.27% for various specimen numbers. [18] uses DWT for rapid material characterization using smart skin for NDT purposes.

In [17], used CWT with CNN to classify defects in narrow overlap weld with visually observable features and reached 96% accuracy. In this work, CWT features are extracted from eddy current testing measurements during welding. The pre-processed resulting 2D scalograms were used as input of CNNs for features extraction and later passed into classifiers. Formerly, similar features were manually extracted and evaluate toward defect identification in friction stir welding [31]. In [20], WTs are employed for automatic inspection of girth weld cracking from Ultrasonic A-scan measures. The authors apply CWT and devise a CNN-SVM model to classify defective and non-defective scans. Moreover, the method is compared with statistical feature extraction, as well as DWT. The most accurate classification is gained by 93.75% on the test set. In [21], fault detection of bearing in rotating machines is automated through contactless thermal imaging and 2D DWT feature extraction. PCA is applied to enhance features, and SVM, LDA, and CDT are used as classifiers, where SVM outperforms the two other methods. In addition, the system was evaluated under three conditions of full-load, half-load, and no-load.

Collecting arc stud welding measurements is a challenging task, as machines are designed to perform accurate weld, rather than collecting comprehensive measurements, and process monitoring. As a result, measurements are inaccessible and attempts for data sampling cause interruptions in production which makes data gathering significantly costly. Hence, to achieve defective welded studs dataset with similar characteristics to production welds, in this chapter, the design of experiments is carefully practiced to form an experimental setup with a novel sensor-based solution and collect a dataset of measurements for various welding conditions leading to defective welded joints.

In addition to data collection, machine learning-based solutions are designed for automation of automotive weld defect classification. In this chapter, two methods based on wavelet transform (WT) are developed and empirically evaluated on the designated MTS dataset of ASW process for feature selection. Afterwards, various 11 types of classifiers including convoloutional neural networks (CNNs) are developed, and utilized for root cause analysis. Furthermore, the effectiveness of synthetic and residual signal generation is investigated. Finally, feature importance is analyzed to extract the most critical measurements in defect root cause analysis, which opens areas for welding parameter optimization.

The remainder of this chapter is organized as follows. Section 8.2 gives an overview of the feature extraction and models. Section 8.3 provides an explanation of data preparation, the dataset, training, evaluation as well as experimental results in detail. Finally, section 8.4 elaborates on model behavior, and in section 8.5 chapter is concluded, as well as future improvements are presented.

8.2 METHODOLOGY

To analyze the root cause of the ASW defects, a set of experiments must be designed to reproduce faulty behavior. During such experiments, various faulty scenarios must be planned to artificially cause defects to occur. In addition, this procedure provides the possibility of analyzing the effect of the weld parameters.

In this work, due to the non-stationary characteristics of weld signals, wavelet transforms are applied to extract features. Initially, preliminary pre-processing and synthetic signal generation is investigated to enhance measurements. Afterward, both CWT and DWT are utilized to extract features. Then a set of classifiers are employed to classify root causes. In order to achieve the best possible accuracy, various mother wavelets are investigated. Additionally, the effectiveness of a data-centric measurement enhancement technique previously presented in [9], is comparatively analyzed. Fig. 8.1 shows the overall pipeline for this research. In the next subsections, the steps are elaborated on in detail.

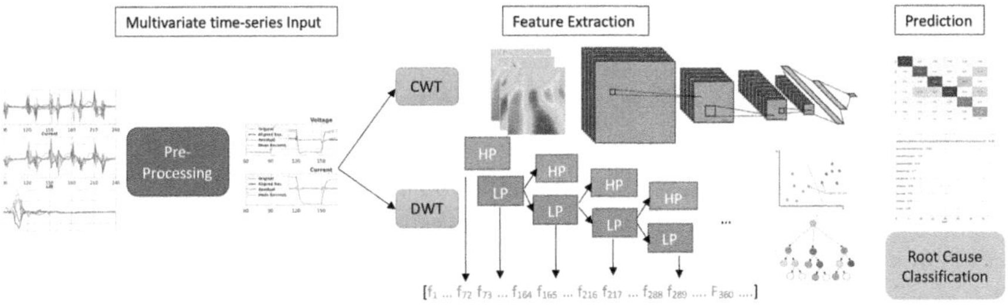

Figure 8.1 Proposed method architecture for root cause analysis of arc stud welds.

8.2.1 NORMALIZATION

Z-score has been used as feature scaling and for quicker convergence of the machine learning models. A sensor measurement is a $T[i]$ sampled with f frequency, and normalization is performed per measurement quantity, e.g. voltage, current, etc. to have a zero mean and unit energy, before passing to the next stage [5] and formulated as:

$$T'[i] = \frac{T[i] - \mu}{\sigma}, \tag{8.1}$$

where μ is the mean of the signals having both the same type of measurement. σ is the standard deviation of the values. This step is the last stage of the data-centric approach, meaning the step is done prior to any feature extraction while it is the last stage of any other measurement enhancement algorithms if applied.

8.2.2 WAVELET TRANSFORMS

In wavelet transform (WT), a series of functions called wavelet is used. These wavelets are derivations from a function called the mother wavelet. Considering ψ, as the mother wavelet, other wavelets will be derived from:

$$\psi_{b,a}(t) = \frac{1}{a^{1/2}} \psi \left(\frac{t-b}{a} \right) \tag{8.2}$$

a, and b, are for transition, and dilation, respectively. Although the wavelet is localized in time, the derivations provide the opportunity to extract both time and frequency information through the multiplication of various wavelets into signal T.

There are many mother wavelets to choose from as the base for feature extraction, and they will be considered as hyper-parameter. Therefore, mother wavelets are usually selected empirically.

There are two distinct approaches for applying these wavelets to the original signal:

8.2.2.1 Continuous Wavelet Transform (CWT)

The Continuous form is formulated as:

$$X_w(a,b) = \frac{1}{(a)^{1/2}} \int_{-\infty}^{\infty} x(t) \bar{\psi} \left(\frac{t-b}{a} \right) dt, \tag{8.3}$$

in which a continuous mother wavelet is applied on signal $x(t)$ and scaling can take infinite values. Using various scales, it is possible to convert time-domain measurements into a 2D scalogram of measurements. These 2D scalograms are time-frequency representations of the measurements, which can provide more details and discriminative features from the signals. A CNN can be ideal to automate the classification of such data.

8.2.2.2 Discrete Wavelet transform (DWT)

For DWT, values assigned to a and b are discretized and can only hold integer values. Similarly, the mother wavelet will be discretized. In the implementation of DWT, the signal gets decomposed by applying a low-pass filter and a high-pass filter, and then the signal is down sampled by a factor of 2 [10] to decompose the signal into lower sub-bands. Fig. 8.1-Feature Extraction visualizes this matter. For each of the lower sub-bands statistical feature extraction, e.g., various percentiles of the signal, mean, etc. is done. Finally, these features will be used as the input of a classifier.

Based on the descriptions provided, two methods of feature extraction approaches are presented. In the following potential classifiers are described.

8.2.3 CLASSIFIER MODELS

CWT represents each measurement as a 2D scalogram, so for each experiment, six 2D scalogram is expected. Therefore, each can be considered as a 6-layer single image and the desired approach is to devise a CNN to classify various root causes. Considering the limited size of the dataset, transfer learning enhances the training and achieves faster and better generalization performance. However, the input shape and size are not compatible with off-the-shelf models, and direct use of off-the-shelf models or transfer learning is not viable. The proposed models are inspired form AlexNet [19] and LeNet5 [14].

Table 8.1, tabulates the proposed CNN architecture. The input layer accepts the $480\times480\times6$ input, where each of the 6 layers is a 2D scalogram of measurement, i.e., CWT extracted coefficients. Then, the first layer is the convolutional layer, which computes the dot product of the 5×5 kernels and kernel regions. This is done by sliding each kernel on each layer's input by performing the convolution operation. The stride is 1 and prior to applying Relu [29] as an activation function, batch normalization is applied, in which inputs are normalized with the mean and variance of the mini-batch. This results in 12 476×476 feature maps. These feature maps will be down-sampled along the height and width using 2×2 max pooling layer and the computing results will be $234\times234\times12$. The same layers (i.e., convolution→batch-norm→Relu→maxpool) will be repeated for 5 times prior to using the Dropout layer. The purpose of the dropout layer is to prevent overfitting, and the dropout probability is set to 0.3. The next step is to flatten the output and pass it into fully connected layers. The flattened output will be mapped into a 1000 neurons dense layer with Relu activation and 6 neurons dense layer with Softmax, respectively.

For DWT features, as mentioned above, statistical extracted features on signals will be fed to various classifiers, including:

- **Naive Bayes [27]**: is a probabilistic supervised machine learning algorithm. The method relies on the Bayes theorem and the independent relationship of features, although the method is relatively fast and works in many real-world scenarios, because of the high dependency on features in our problem, it is not expected to outperform more complex methods.

Table 8.1
The applied CNN Architecture.

Feature Name	Output Shape
Conv 2D	(476,476,12)
Batch Normalization	(476,476,12)
Max pooling	(238,238,12)
Conv 2D	(234,234,16)
Batch Normalization	(234,234,16)
Max pooling	(117,117,16)
Conv 2D	(113,113,32)
Batch Normalization	(113,113,32)
Max pooling	(56,56,32)
Conv 2D	(52,52,32)
Batch Normalization	(52,52,32)
Max pooling	(26,26,32)
Conv 2D	(22,22,32)
Batch Normalization	(22,22,32)
Max pooling	(11,11,32)
Dropout	(11,11,32)
Flatten	(3872)
Dense	(1000)
Dense	(6)

- **Gradient Boosting [28]**: in general boosting refers to the idea of modifying a weak classifier to boost its performance. gradient boosting is an ensemble method, and instead of fitting a single classifier, it fits a new classifier on the error of the previous classifier.
- **AdaBoost**[15]: Similar to other boosting algorithms, AdaBoost adaptively ensemble classifiers to build a strong one. It leverages decision stumps, which use a single variable. Adaboost assigns larger weights to both incorrectly classified samples and high accuracy trees. Thus, in the next iterations, they will be re-attempted with higher priority through stronger classifiers.
- **Random Forest (RF)**: is an ensemble learning method and originated from gradient boosting [12]. RF consists of several independent Decision Trees. In RF, different subsets of the original training set are generated using a bootstrap sampling method, and each independent tree is trained separately. All the trees together can generate a correct result, while there is a high probability of producing incorrect results by a single tree. In fact, the result of RF is generated by averaging or voting all DTs.

Figure 8.2 The experiment setup. From left: Tucker SWS control unit and stud feeder; robotic arm which performs the contact with workpiece and conducts welding; robotic arm end effector and the weldhead; workpiece, stud and welded studs samples.

- **XGBoost**: Extreme Gradient Boosting is also designed on top of DTs. It is initially introduced in [11], and utilizes an iterative manner to merge DTs to boost the performance of the classifier. In fact, each DT depends on the previous DT to develop a better tree. Overfitting is also avoided using regularization terms like L1 and L2.

8.3 EXPERIMENTS

8.3.1 DATASET AND DESIGN OF EXPERIMENT

In this work, ASW is automated through a robotic arm and is performed for the following experiments, as shown in Fig. 8.3. Fig. 8.2 visualizes the experimental setup. Stud welding system (SWS), Tucker TE 1500, is used for welding, as well as TH 510 for the weldhead. Automation is conducted through FANUC Robot R-200iB. Various parameters such as voltage, current, linear motor (LM) current, fastener positioning, and their reference signals are monitored and recorded through sensor readings during the operation. Although some measurement seems less relevant, they are all involved in the process and have contributed to the quality of the final product. For example, current is considered as a controlling parameter. However, with removal of current signal, regression in model performance is observable. Therefore, it is considered for evaluations. Similar to [9], for these experiments, average penetration is 0.70, and the sampling duration is 500ms. The sampling rate is approximately 1kHz, and finally, all logged sensors' data are sent through the MQTT protocol to the processing unit.

In the experiments, a set of synthetic defects or root causes are generated based on the design of the experiment principles. Four types of pollution on the welding surface are made and coded as root cause (RC) 1 to RC 4. Moreover, two other types are added on stud quality and electrical system; and coded into RC 5 and RC 6.

Table 8.2
Class Distribution for Dataset.

Classes	*RC1*	*RC2*	*RC3*	*RC4*	*RC5*	*RC6*
Count	59	60	54	48	18	18

The initial dataset contains six classes of Arc Stud Welds with various welding conditions, as mentioned, and measurement types include Voltage, Current, LM Current, and its reference, lift position, and its reference sums up to 6 measurements type. Moreover, depending on the experiment, some of such measurements will undergo alteration, which will be explained under appropriate subsection. Finally, the dataset has 257 samples from 6 classes (RC1, ..., RC6), the distribution of which is shown in Table 8.2. Each instance has six measurements with 480 timestamps including Voltage, Current, Penetration, LMCurrent and its reference, lift position and its reference.

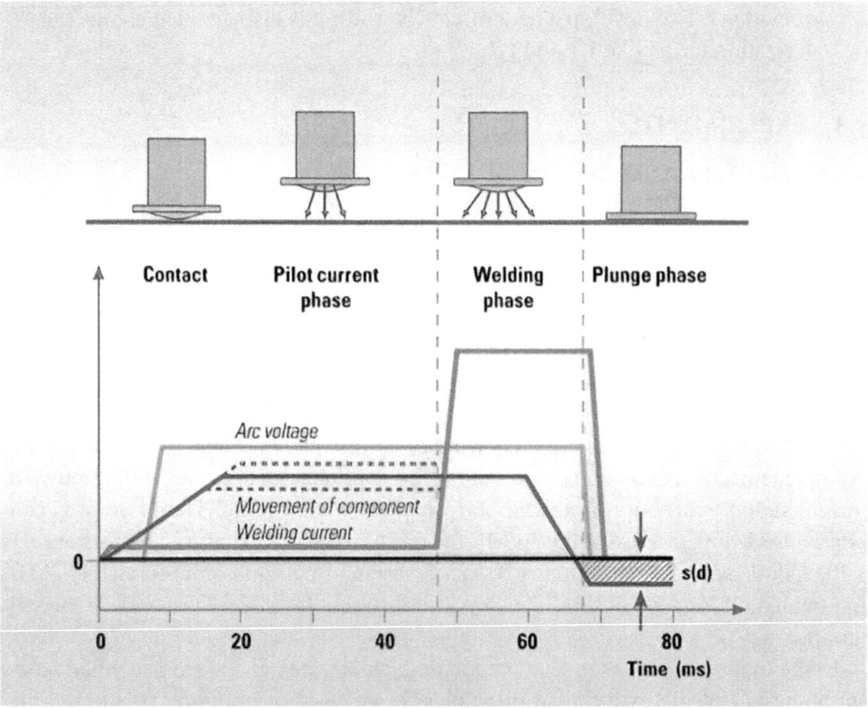

Figure 8.3 Arc stud welding process [22].

8.3.2 SYNTHETIC MEASUREMENT GENERATION

In this chapter, in addition to original dataset, synthetic and residual measurement generation are taken into account to help feature extraction and transformation. The approach results in transforming the original dataset, which is referred to residue dataset [9]. This is performed prior to wavelet transformation, and once done, identical WTs and models will be trained and tested on both original, and residue dataset.

The main goal of the synthetic measurement generation is to calculate signal residuals or their existing references, if they exist, to augment discriminative features in the signals. Algorithm 2 summarizes the steps for synthetic measurement generation. For each measurement, if the reference signal exists, the residual calculation would be the subtraction reference from the original measurement (e.g., LM current). However, for measurements that reference is not available, a mean of the measurements is calculated through independent component analysis (ICA) [6] transform followed by inverse ICA. As a result, a mean approximation of the signal is generated and will be used as the reference for the measurements. The final step prior to subtraction would be signal alignment through dynamic time warping (DTW) [7] to address the shifts in signal in comparison with ICA approximation. A comprehensive explanation is presented in [9].

Algorithm 2: Synthetic measurement and residual generation algorithm.

```
Result: residuals of T' as residuals_set ;
residuals_set ← [ ] ;
if reference signal (T̂) exists then
    while not processed every instance T do
        T' ← T − T̂;
        residuals_set ← T' ;
    end
else
    ICA,T̂_mica ← [] ;
    ICA ← fit ICA for entire set of of Ts ;
    ICA ← Inverse Transform ICA ;
    T̂_mica ← calculate mean of reconstructed signals over x-axis;
    while not processed every instance T do
        T ← DTW(T, T̂_mica) ;
        T' ← T − T̂_mica;
        residuals_set ← T' ;
    end
end
normalize residuals_set using (8.1);
```

Table 8.3
statistical features that are used to perform feature extraction from signals. N is the length of the signal, and Med is the median function.

Feature Name	Formula
Entropy	$-\sum_{i=1}^{N} p_i \log_2 p_i$
n^{th} Percentile	$\lceil \frac{n}{100} \times N \rceil$
Median	$Med(x)$
Mean	$M = \frac{1}{N} \sum_{i=i}^{N} x_i$
STD	$\sqrt{\frac{1}{N-1} \sum_{i=1}^{N} (x_i - M)^2}$
Variance	$\sqrt{\frac{1}{N} \sum_{i=i}^{N} (x_i - M)^2}$
Root Mean Square	$\sqrt{\frac{1}{N} \sum_{i=i}^{N} (x_i)^2}$
Zero Crossing	$\sum_{i=2}^{N} \lvert sign(x_i) - sign(x_{i-1}) \rvert$
Mean Crossing	$\sum_{i=2}^{N} \lvert sign(x_i - M) - sign(x_{i-1} - M) \rvert$

8.3.3 CWT

As described in 8.2.2.1, a mother wavelet function is utilized to generate wavelets through (8.2), and followed by (8.3), it is possible to generate 2D scalograms of the signals. In terms of scales, the common practice is to use an equal number to signal length, which is 480, and results in an squered shape scalogram. For implementation, Pywavelet [13], is employed. Fig. 8.4 visualizes 2D scalograms for measurements related to a single instance, in which Morelet is used as the mother wavelet.

8.3.4 DWT

As Fig. 8.1-(Feature Extraction) and section 8.2.2.2 illustrate, original measurements will be decomposed into sub-bands, and on each sub-band feature extraction is performed. Max decomposition level is calculated through:

$$max\_level = \left\lfloor \log_2 \left(\frac{sample\_len}{moder\_wavelet\_len - 1} \right) \right\rfloor \tag{8.4}$$

meaning with different mother wavelets, the depth of decomposition is different and ranges from 3 to 8 levels. Finally, statistical feature extraction is done on extracted signal coefficients and per measurement. Hence, from each level, features in Table 8.3 are extracted and concatenated. As a result, for each level, there will be 72 values (i.e., 12 features and 6 measurements). For instance, if the *max_level* is 8, 576 features exist for a single sample. Note that for n^{th} percentile feature, 5th, 25th, 75th, and 95th percentiles are extracted.

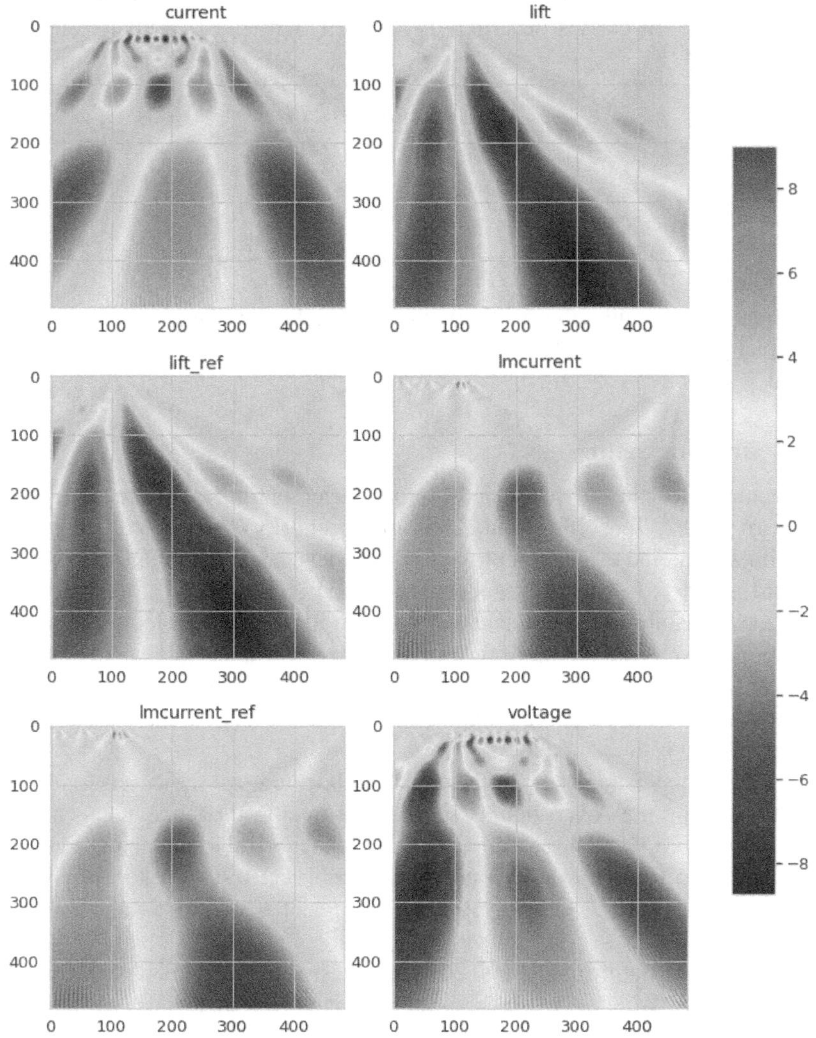

Figure 8.4 2D scalograms are generated using CWT. Each measurement generated a 480×480 feature map, and a single instance is described by a 480×480×6 data.

8.3.5 MODELS

8.3.5.1 Evaluation Metrics

The performance of the models in this chapter is evaluated through the F1-score macro metric. F1-score macro is the mean of per-class F1-score, and that is:

$$F1 = \frac{1}{C}\sum_{i=i}^{C} F1_i \tag{8.5}$$

where C is the number of classes, and $F1_i$ is the F1-score for class i. In this way, low performance on classes penalizes the final F1 harder. $F1_i$ itself is calculated using:

$$F1_{\text{i}} = 2 \cdot \frac{precision_{\text{i}} \cdot recall_{\text{i}}}{precision_{\text{i}} + recall_{\text{i}}} \tag{8.6}$$

where,

$$precision_{\text{i}} = \frac{tp_{\text{i}}}{tp_{\text{i}} + fp_{\text{i}}}; recall_{\text{i}} = \frac{tp_{\text{i}}}{tp_{\text{i}} + fn_{\text{i}}} \tag{8.7}$$

in which tp_{i} is the count of correctly predicted samples for class i, fp_{i} is the count of incorrect predictions for class i, and fn_{i} is the count of predictions that incorrectly assigned as class i.

8.3.5.2 Training Setting

Implementation of the models and hyperparameter optimization are done using scikit-learn v1.0.2, Tensorflow and Keras v2.13, and PyWavelets v1.3. For training, optimization, and evaluation of the CNN models, a single NVIDIA GTX 1080 GPU card is utilized. All DWT models are trained on CPU Intel Core i7 CPU @ 3.2 GHz. All feature extraction and testing with all 106 pre-implemented mother wavelets and 10 models are performed in a one and half-hour training session with parallel processing. However, for CNN-based models, both feature extraction and model training were timely, regarding the large size of images, and the number of parameters in CNN deep models.

The 3.9 million parameters CNN model architecture as shown on Table 8.1, is examined using the Adam optimizer (learning rate: 0.01) across 8 epochs with early stopping on validation set accuracy. A cross-categorical loss function addresses multi-class classification. For XGBoost, after empirical tests, the top performing model is trained with 0.3 learning rate, max depth of 6, and 100 estimators with 1 sub sample.

The dataset is stratified split into train, dev, test sets with ratios of 60%, 20%, 20%, respectively. We used 5-fold cross validation on the train and dev sets, then the performance of the best of models are reported on the test set.

8.3.6 CNN MODELS

As shown in Fig. 8.4, there are 6 types of measurement, and for each measurement, a 2D scalogram is generated with size (480×480); Morlet mother wavelet is used for feature extraction. By concatenating these measurements, each instance will have the shape (480,480,6). Thus, even with resizing the input to a smaller shape, an off-the-shelf model would not accept such input. Therefore, transfer learning is not directly possible, and it is required to re-architect existing architectures. While appending additional input layer might help mitigating transfer learning issue with the cost of

Table 8.4
Performance of the models with various approaches. *sg* **stands for synthetic generation that is performed prior to wavelet transforms.**

Method	Model	Input size	Mother wavelet	mean val F1 score	F1 score
CWT	proposed CNN	$480^2 \times 6$	morl	62.1	61.61
CWT+*sg*	proposed CNN	$480^2 \times 6$	morl	64.5	63.66
DWT	GradBoosting	360	rbio3.9	85.8	84.96
DWT	XGBoost	360	sym13	85.8	85.41
DWT+*sg*	XGBoost	288	DB17	84.6	84.04
DWT+*sg*	XGBoost	432	bior3.5	83.3	83.62

re-training additional layers, this also experimentally did not address the problem in this case.

Since the dataset is relatively small, very deep models like ResNet50, even with regularization, have a great potential for overfitting. Therefore, well-known models with smaller number of layers like LeNet5, and VGG16 are utilized as the basis to re-architect and train. The performance of the models is depicted in Table 8.4.

8.3.7 DWT CLASSIFIERS

As described in 8.2.2.2, the input of each classifier is the *(level of decomposition+1)* $\times$ 12 (i.e., number of features extracted) $\times$ 6, i.e., number of measurements. In Py-wavelet, there are 106 pre-implemented discrete mother wavelets to apply for feature extraction.

Fig. 8.7 demonstrates the performance of 10 classifiers on features extracted on basis of 106 various mother wavelets, with F1 score. For performances reported in Fig. 8.7-(a), they are enhanced with Algorithm 2, and normalized by 8.1, while Fig. 8.7-b demonstrates results related to normalized raw data processed with DWT algorithm for feature extraction.

Generally, ensemble classifiers perform better for the task. The best model reaches an accuracy of 85.41% on the test set, with Sym13 as the mother wavelet, and XG-Boost as the classifier, with original dataset. The confusion matrix for per class performance for this model is visualized in Fig. 8.5. The poorest performance is related to RC6, and RC5 which is related to impurity in stud.

While modeling, hyperparameters, and mother wavelets play a key role in the model performance, extracted statistical features can have a major or minor role as well. For example, Crossings and Entropy play a major role. Dropping each leads to 10-20% drop in the performance of the state-of-the-art model. In contrast, n^{th} percentiles has a less significant impact of 1-3% drop in accuracy. Moreover, addition of extra n^{th} features, ends up in either no impact or minor drops in the metrics.

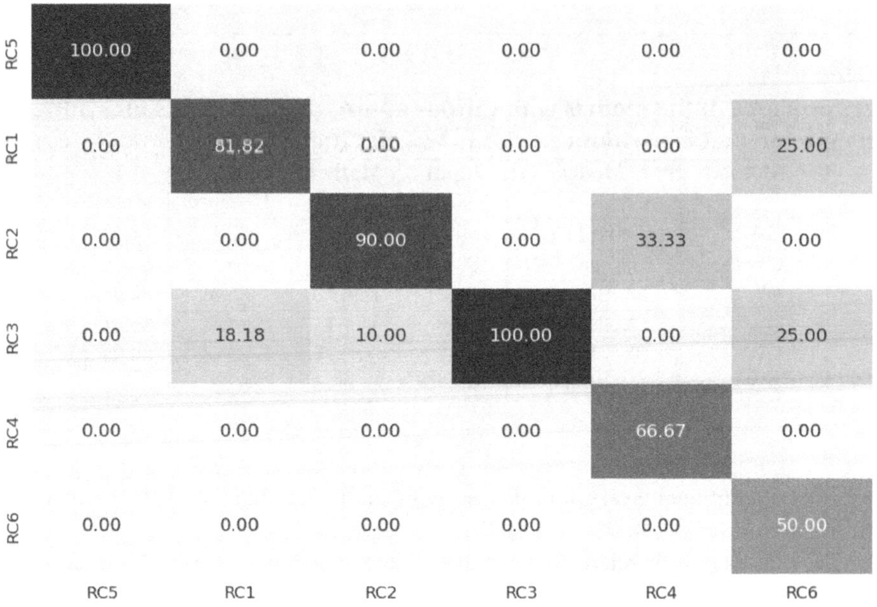

Figure 8.5 Confusion Matrix of best model based on DWT features and XGBoost classifier. The values are normalized according to predictions.

Fig. 8.6 demonstrates the key features (model gain) related to the top-performing XGBoost classifier. It is demonstrated that voltage and lift changes have a key role in the classification of the root cause. In addition, low frequencies (L3 to L5) are essential for classification.

8.4 DISCUSSION

As evaluations reveal, both CWT and DWT are used for feature extraction. In general, DWT-based classifier models outperform CWT and CNNs. An explanation for this is that CNNs are data-hungry. On the one hand, because of the rigorous nature of collecting annotated faulty data for these experiments and cost of experiments, the dataset was small in compare with richness of data for each instance. On the other hand, the 6 measurements required the model to accept input with 6 layers (i.e., 480×480×6), hence, benefiting from pre-trained off-the-shelf CNN classifiers for transfer learning becomes challenging in this context. These two reasons including the fact that the 6 layers contain visually imperceptible features for model to detect, explain the low performance and high latency of the CNN models. In addition, deeper models quickly overfit due to the size of the dataset.

For DWT features, 106 pre-implemented mother wavelets and 10 classifiers are investigated. In addition, all the experiments are performed both with original and residue datasets 2. The best model has reached F1-score of 85.41%, and it is achieved

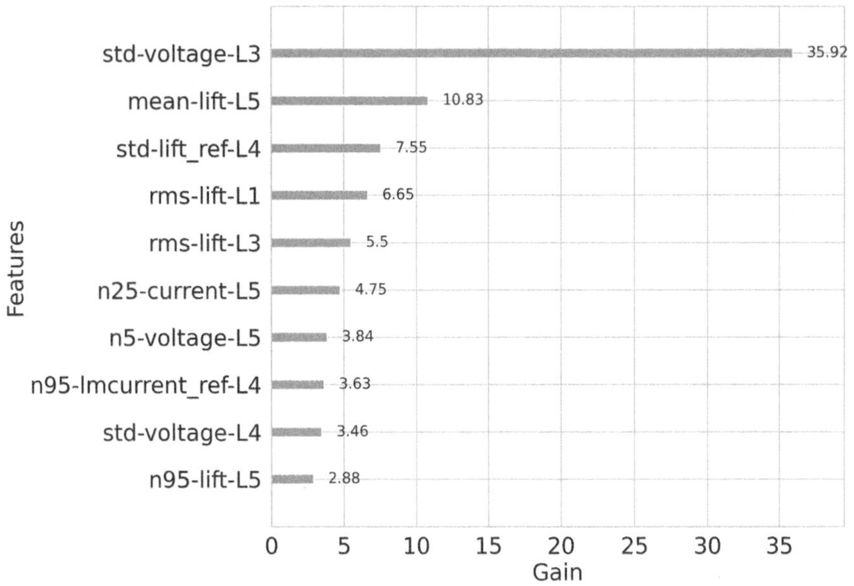

Figure 8.6 Feature Importance (Gain) based on DWT features and XGBoost classifier.

with original dataset. Although the best model is on the dataset with no synthetic measurement generation, by comparing Fig. 8.7-(a) (no synthetic generation) and Fig. 8.7-(b) (with synthetic generation), it is discernible that synthetic residual data has made features more distinct hence all classifiers reach 20 to 30 percent better performance.

In terms of mother wavelets, there is no specific pattern for any family of mother wavelets. Fig. 8.7, illustrates that while a specific mother wavelet, e.g., Coif15, might show great performance with a specific classifier, i.e., XGBoost, it does not necessarily shows the same high performance when compared to other classifiers. Thus, the mother wavelet should be considered as a hyper-parameter and should be tuned with respect to the classifier.

8.5 CONCLUSION

In this chapter, a new approach for root cause analysis and defect classification of arc stud welding is presented. The approach is empirically and comparatively evaluated on a dataset of multivariate time series measurements collected during the welding process. The proposed solution employs continuous and discrete wavelet transforms for feature extraction. In addition, the effectiveness of the synthetic and residual measurement generation strategy as well as the feature importance of the best classifier are comparatively analyzed. The best model utilized DWT features and XGBoost for

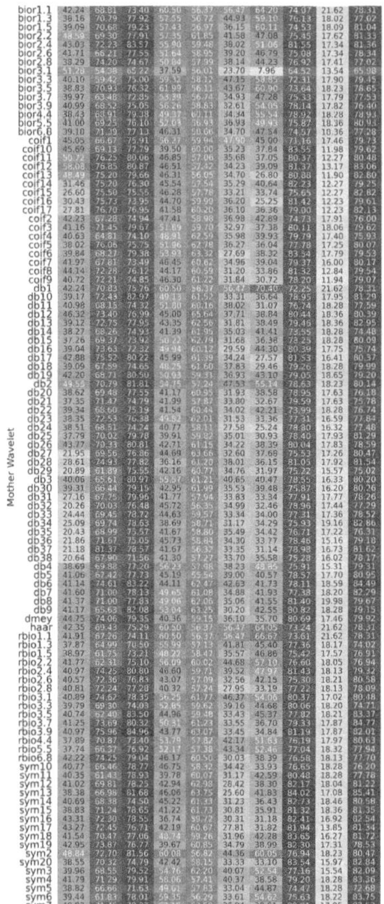

(a) DWT classifiers on original dataset

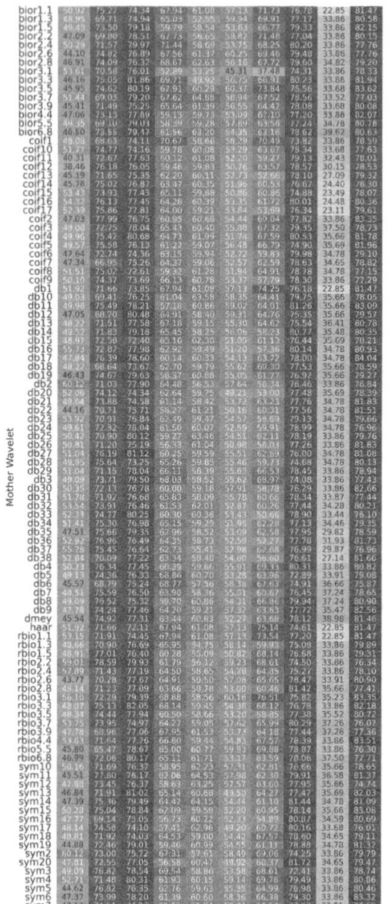

(b) DWT classifiers on residue dataset

Figure 8.7 Performance of DWT classifiers in F1-score

the classification of root causes and reached F1-score of 85.41%. Based on feature importance, low-frequency voltage measurements and standard deviations from lift position have the highest gain in XGBoost classification.

Although the work shows impressive results in root cause detection as well as real-time performance, the work is in the research and development stage indicating requirement of further field evaluation. As an iteration on this research one can consider empirically investigating effectiveness of various statistical feature extraction methods of DWT performance. As another area and in practical implementation, one of the challenges is optimization and domain adaptation the work toward in production line deployment. Extensive investigation of unsupervised approaches is another interesting area.

REFERENCES

1. Ivan Samardzic, Stefanija Klaric, and Thomas Siewert. Analysis of Welding Parameter Distribution in Stud Arc Welding. In *Proceedings of the Conference: Welding and Materials*, Dubrovnik, Croatia, July 2007.
2. Nabeel K Abid Al-Sahib, Hussam K Abdul Ameer, and Saif Ghazy Faisal Ibrahim. Monitoring and quality control of stud welding. *Al-Khwarizmi Engineering Journal*, 5(1):53–70, 2009.
3. Harry A Chambers. Principles and practices of stud welding. *Pci Journal*, 46(5):46–59, 2001.
4. Hassan Ismail Fawaz, Germain Forestier, Jonathan Weber, Lhassane Idoumghar, and Pierre-Alain Muller. Deep learning for time series classification: a review. *Data Mining and Knowledge Discovery*, 33(4):917–963, 2019.
5. Dina Q Goldin and Paris C Kanellakis. On similarity queries for time-series data: constraint specification and implementation. In *International Conference on Principles and Practice of Constraint Programming*, pages 137–153. Springer, 1995.
6. Kuan-Heng Chen and Khaldoun Khashanah. The Reconstruction of Financial Signals Using Fast ICA for Systemic Risk. In *2015 IEEE Symposium Series on Computational Intelligence*, pages 885–889, 2015.
7. Tomasz Górecki and Maciej Łuczak. Multivariate time series classification with parametric derivative dynamic time warping. *Expert Systems with Applications*, 42(5):2305–2312, 2015.
8. Zhiguang Wang, Weizhong Yan, and Tim Oates. Time series classification from scratch with deep neural networks: A strong baseline. In *2017 International joint conference on neural networks (IJCNN)*, pages 1578–1585. IEEE, 2017.
9. Sadra Naddaf Shargh, Mahdi Naddaf-Sh, Maxim Dalton, Soodabeh Ramezani, Amir R Kashani, and Hassan Zargarzadeh. Explainable Models for Multivariate Time-series Defect Classification of Arc Stud Welding. *International Journal of Prognostics and Health Management*, 14(3), 2023.
10. Ying-Yi Hong and Mark Tristan Angelo M. Cabatac. Fault Detection, Classification, and Location by Static Switch in Microgrids Using Wavelet Transform and Taguchi-Based Artificial Neural Network. *IEEE Systems Journal*, 14(2):2725–2735, 2020.
11. Tianqi Chen and Carlos Guestrin. XGBoost: A Scalable Tree Boosting System. In *Proceedings of the 22nd ACM SIGKDD International Conference on Knowledge Discovery and Data Mining*, pages 785–794, San Francisco, California, USA, 2016. ACM.
12. Leo Breiman. Random forests. *Machine learning*, 45(1):5–32, 2001.
13. Gregory Lee, Ralf Gommers, Filip Waselewski, Kai Wohlfahrt, and Aaron O'Leary. PyWavelets: A Python package for wavelet analysis. *Journal of Open Source Software*, 4(36):1237, 2019.

14. Yann LeCun, Léon Bottou, Yoshua Bengio, and Patrick Haffner. Gradient-based learning applied to document recognition. *Proceedings of the IEEE*, 86(11):2278–2324, 1998.
15. Yoav Freund and Robert E Schapire. A decision-theoretic generalization of on-line learning and an application to boosting. *Journal of computer and system sciences*, 55(1):119–139, 1997.
16. Ehab N. Abbas, Salman Omran, Muhaed Alali, Mohammed H. Abass, and Adnan N. Abood. Dissimilar Welding of AISI 309 Stainless Steel to AISI 1020 Carbon Steel Using Arc Stud Welding. In *2018 International Conference on Advanced Science and Engineering (ICOASE)*, pages 462–467, 2018.
17. Rui Miao, Yuntian Gao, Liang Ge, Zihang Jiang, and Jie Zhang. Online defect recognition of narrow overlap weld based on two-stage recognition model combining continuous wavelet transform and convolutional neural network. *Computers in Industry*, 112:103115, 2019.
18. Rajendra Prasath Palanisamy, Subrata Mukherjee, Mahmood Haq, and Yiming Deng. Rapid Material Characterization using Smart Skin with functional Data Analysis. In *PHM Society European Conference*, volume 6, pages 7–7, 2021.
19. Alex Krizhevsky, Ilya Sutskever, and Geoffrey E Hinton. Imagenet classification with deep convolutional neural networks. *Advances in neural information processing systems*, 25, 2012.
20. Y Yan, D Liu, B Gao, GY Tian, and ZC Cai. A deep learning-based ultrasonic pattern recognition method for inspecting girth weld cracking of gas pipeline. *IEEE Sensors Journal*, 20(14):7997–8006, 2020.
21. Anurag Choudhary, Deepam Goyal, and Shimi Sudha Letha. Infrared thermography-based fault diagnosis of induction motor bearings using machine learning. *IEEE Sensors Journal*, 21(2):1727–1734, 2020.
22. STANLEY Engineered Fastening. Stud Welding Products & Systems, 2021. `http://www.emhart.eu/eu-en/products-services/products-by-category/tucker-stud-welding/stud-welding-systems/energy-units.php`. Accessed: 05-26-21.
23. PB Garcia-Allende, J Mirapeix, OM Conde, A Cobo, and JM Lopez-Higuera. Spectral processing technique based on feature selection and artificial neural networks for arc-welding quality monitoring. *Ndt & E International*, 42(1):56–63, 2009.
24. Jun Young Jeon, Duhwan Kim, Gyuhae Park, Eric Flynn, To Kang, and Soonwoo Han. 2D-wavelet wavenumber filtering for structural damage detection using full steady-state wavefield laser scanning. *NDT & E International*, 116:102343, 2020.
25. Ki-Bok Kim, David K Hsu, and Daniel J Barnard. Estimation of porosity content of composite materials by applying discrete wavelet transform to ultrasonic backscattered signal. *NDT & E International*, 56:10–16, 2013.
26. Juan Dong, Guocheng Xu, Hongyang Yu, Guoquan Fan, Lingbo Wei, and Xiaopeng Gu. Connection status evaluation in arc stud weld joints by ultrasonic detection. *The International Journal of Advanced Manufacturing Technology*, 100(1):663–672, 2019.
27. David J Hand and Keming Yu. Idiot's Bayes—not so stupid after all? *International statistical review*, 69(3):385–398, 2001.
28. Jerome H Friedman. Greedy function approximation: a gradient boosting machine. *Annals of statistics*, pages 1189–1232, 2001.
29. Kunihiko Fukushima. Cognitron: A self-organizing multilayered neural network. *Biological cybernetics*, 20(3-4):121–136, 1975.

30. Tao Liang, Wenwei Ren, Gui Yun Tian, Mutaz Elradi, and Yunlai Gao. Low energy impact damage detection in CFRP using eddy current pulsed thermography. *Composite Structures*, 143:352–361, 2016.
31. Shilpi Kumari, Rahul Jain, Ujjwal Kumar, Inderjeet Yadav, Nitin Ranjan, Kanchan Kumari, Ram Kumar Kesharwani, Sachin Kumar, Srikanta Pal, Surjya K Pal, and others. Defect identification in friction stir welding using continuous wavelet transform. *Journal of Intelligent Manufacturing*, 30:483–494, 2019.

Section IV

Ultrasonic Testing

9 AI-Driven Weld Inspection with Ultrasonic B-Scans

Amir-M. Naddaf-Sh, Vinay S Baburao, Hassan Zargarzadeh

CHAPTER SUMMARY

Automated ultrasonic testing (AUT) is a non-destructive testing (NDT) method widely employed in industries that hold substantial economic importance. To ensure accurate inspections of exclusive AUT data, expert operators invest considerable effort and time. While artificial intelligence (AI)-assisted tools, utilizing deep learning models trained on extensive in-laboratory B-scan images, whether they are augmented or synthetically generated, have demonstrated promising performance for automated ultrasonic interpretation, ongoing efforts are needed to enhance their accuracy and applicability. This is possible through the evaluation of their performance with experimental ultrasonic data. In this study, we introduced a real-world ultrasonic B-scan image dataset generated from proprietary recorded AUT data during industrial automated girth weld inspection in oil and gas pipelines. The goal of inspection in our dataset was detecting a common type of defect called lack of fusion (LOF). We experimentally evaluated deep learning models for automatic weld defect detection using this dataset. Our assessment covers the baseline performance of state-of-the-art (SOTA) models, including transformer-based models (DETR and Deformable DETR) and YOLOv8. Their flaw detection performance in ultrasonic B-scan images has not been reported before. The results show that, without heavy augmentations or architecture customization, YOLOv8 outperforms the other models with an F1 score of 0.814 on our test set.

9.1 INTRODUCTION

Ultrasonic testing (UT) is a non-destructive testing (NDT) technique used for inspecting components and structures to ensure their integrity and quality. The advantage of employing NDT methods is that the structures can be inspected without interruptions in their daily functions, offering a nonintrusive means of assessing structural health. NDT is widely used in safety-critical industries such as aerospace, automotive, and oil and gas, where the reliability of structures plays an important role. To enhance the overall confidence and efficiency of the inspection process, researchers are increasingly exploring new techniques such as optimizing laser-induced arrays for defect detection [1], implementing automated computer-aided inspection for real-time defect detection in laser-welded blanks [2], etc. In comparison to other NDT methods, UT boasts a long history of effectively detecting subsurface [3] and small

DOI: 10.1201/9781003766575-9

flaws [4]. In UT, a transducer placed on the surface emits ultrasound waves through the material. By analyzing the waves reflected back to the transducer, corrosion and flaws that occurred during manufacturing, welding, or other processes can be detected and assessed.

Automated ultrasonic testing (AUT) utilizes automated systems mounted around the structure, performing precise and rapid inspection with phased array probes. In phased array technology, each element in the array is pulsed and delayed independently, creating a varied range of beam angles and focal points for the comprehensive examination of materials. Presently, AUT inspection entails an analysis of an ever-expanding volume of acquired data by human operators. The responsibilities during AUT inspections encompass a range of tasks for operators. These include positioning the inspection head around the structure, conducting inspections, interpreting data, recording results, and making decisions as to whether to accept or reject the scanned profiles. These intricate, sequential tasks, coupled with the multitude of scans requiring inspection, could compromise accuracy and extend the time needed for UT interpretations.

In UT, data are often recorded in the form of A-scans, B-scans, and C-scans. B-scans, which visualize more detailed information about the inspected material, have captured considerable interest within the field of UT. After the introduction of ImageNet [5] in the computer vision community, substantial efforts have been directed toward harnessing deep-learning-based algorithms to tackle challenges in downstream tasks, especially in the interpretation of ultrasonic B-scan images.

In [6], the authors conducted experiments using both conventional algorithms, such as flaw classifiers based on hand-crafted features and statistical techniques, and machine-learning-based methods, including convolutional neural networks (CNNs), to classify defects in images obtained from the laser-generated imaging of ultrasonic wave propagation [7, 8] on stainless-steel plates. In their subsequent work [9], first, they released the first publicly available dataset for ultrasonic inspection named USimgAIST [10], consisting of 7004 images with both normal cases and defective cases from 18 stainless steel plates including drill holes and slits flaws. Second, they trained multiple SOTA deep CNNs on their dataset to benchmark deep learning models for automatic ultrasonic image interpretation. In [11], a reinforcement-learning-based neural architecture search neural network (RL-NAS NN) was utilized to automatically implement the optimal CNN-based architecture for classifying defects on the USimgAIST dataset. All previous methods, despite proposing a novel idea for ultrasonic image interpretation, suffer from an inability to detect the size and the location of the defects.

The first application of SOTA CNN-based object detectors for AUT was implemented in [12]. They fine-tuned YOLOv3 [13] and SSD [14] on their dataset containing 490 grayscale B-scan images with 1562 annotations using heavy augmentation methods. Another SOTA one-stage object detector, EfficientDet-D0 [15], using custom-calculated anchors, was exploited in [16] to detect flaws on their in-house dataset containing 4147 B-scan images that were artificially flawed. DefectDet [17], a novel deep learning architecture, was created by replacing the default backbone of

EfficientDet with a lightweight encoder–decoder-based feature extractor and a custom detection head for detecting objects with extreme aspect ratios by shifting the input to the biFPN block in the architecture. Despite the advancements in automated defect detection techniques, the reliability of these deep learning models heavily relies on a large and well-distributed dataset.

Generating a substantial amount of ultrasonic B-scan data is time-consuming and costly. Therefore, many studies on classifying and localizing flaws in ultrasonic B-scan images have often relied on in-laboratory images from inspected test specimens under controlled conditions or a limited amount of real-world UT image data that are multiplied using techniques such as image augmentation or synthetic data generation to supply data-hungry deep neural networks (DNNs) for training. Moreover, to build a dataset with a variety of flaws, researchers have commonly introduced artificial defects into images or have intentionally created flaws on test specimens.

In [18], the authors employed extensive data augmentation for their virtually flawed ultrasonic data [19] with varying backgrounds to create their 10,000-size image dataset. With the advent of AI, instead of relying solely on augmentation, more reliable methods for extending the number of training inputs have emerged. In [20], the authors proposed two generative-adversarial-based networks, Detection-GAN [21], and a modified SPADE GAN [22] for generating synthetic B-scan images that are hardly indistinguishable from the real ones for UT operators. Despite the significant advancements made in the field of AUT, the lack of experiments conducted on datasets derived from extreme environments is apparent.

It is evident that the distribution of data obtained from in-laboratory experiments may differ from that of onshore inspections. This discrepancy could pose a significant concern for applications where deep-learning-based models play a role in the decision-making process for ultrasonic interpretation. In this chapter, we introduce a proprietary B-scan image dataset that we generated from recorded reports of onshore automated girth weld inspections of oil and gas pipelines in extreme conditions acquired by UT experts. Additionally, the type of flaw present in our dataset is a lack of fusion (LOF), a common flaw in automated girth welding. We believe that this is the first time in the literature that ultrasonic B-scan images captured from industrial environments with real flaws, specifically LOF, are utilized for fine-tuning deep learning algorithms. We aimed to investigate the performance of SOTA deep-learning-based models without any additional synthetic data or modification to the architectures to see their baseline performance on industrial B-scan images. We illustrate our work in Figure 9.1, where an expert human operator performs automated ultrasonic inspections onshore. Once the inspection is completed, the data are captured and stored using acquisition software. Subsequently, the inspected data are processed and fed into a neural-network-based flaw detector, which analyzes the data to identify potential flaws. Finally, the expert human operator reviews and validates the results, providing a crucial check to correct any missed or incorrect detections.

The contributions of this work are as follows:

- We introduce a proprietary industrial ultrasonic B-scan images dataset that is generated from recorded AUT inspections performed by human inspectors during onshore oil and gas girth weld inspections.

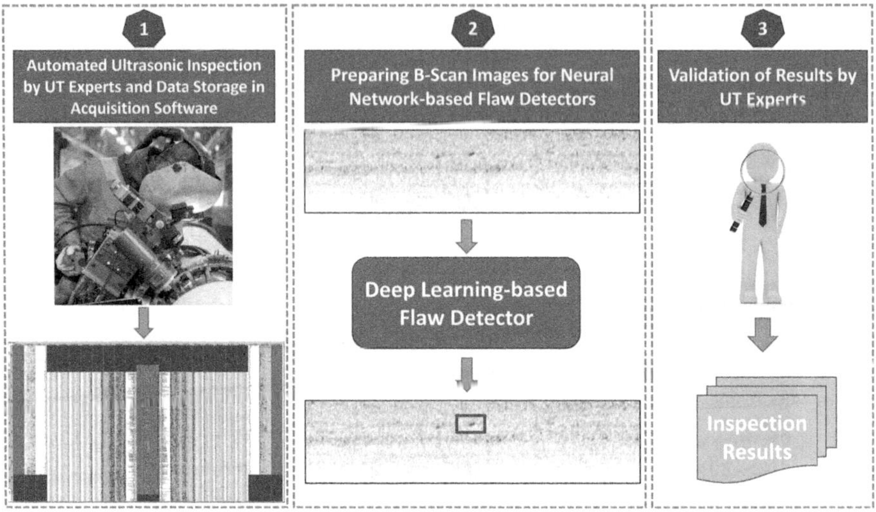

Figure 9.1 Quick summary of our study.

- In addition to the B-scan dataset, genuine flaws that usually occur during automated girth welding, such as LOF, are introduced and investigated.
- We fine-tune and evaluate SOTA models, such as transformer-based architectures (DETR and Deformable DETR) and YOLOv8, on industrial ultrasonic B-scan images.

The remainder of this article is organized as follows: Section 9.2 provides details on dataset preparation and properties. Section 9.3 explains the implementation, training, and evaluation processes of several experiments. Section 9.4 compares the performance of several SOTA models on our B-scan dataset. Finally, Section 9.5 concludes this chapter and presents our work's limitations and future studies.

9.2 DATASET

9.2.1 DATASET DESCRIPTION

We have created a dataset of Ultrasonic B-scan images from weld inspection data acquired by UT experts with automated girth weld inspection system utilizing phased array technology for inspecting oil and gas pipeline. The inspection system utilizes the zone discrimination technique for weld inspection. In this approach, each weld is divided into a few vertical zones, each having a specific beam for inspection. Multiple beams travel around the pipe in a single linear movement. Zone discrimination divides the weld into three main regions: The Cap region, Body region, and Root region. The Cap and Body regions include zones such as Fill 1 and Fill 2, respectively. The Root region includes Hot pass 2, Hot pass 1, LCP, and Root. Figure 9.2

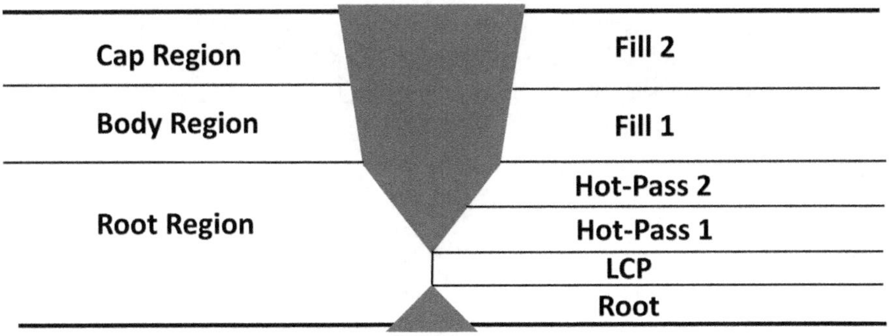

Figure 9.2 Weld Zones.

represents the three regions in the Zone discrimination technique. In our dataset, the weld type is J-bevels, and the weld defect class is lack of fusion (LOF), which is the most common defect in automated girth welding.

9.2.2 DATASET PREPARATION

We created our dataset using 87 proprietary weld inspection records recorded by UT experts with the inspection system during onshore oil and gas pipeline girth weld inspections. We were not involved in the inspection process; we only collected B-scan images from the proprietary weld inspection data provided to us by UT experts. Each recorded weld datum includes a 1920 × 1080 image containing multiple strip charts representing the output of the channels of the phased array probes used for AUT. The strip charts also provide views of the downstream and upstream phased array probes, as well as each weld zone that each channel represents. Moreover, each recorded weld datum contains a table filled out by human operators including the type, size, and location of the detected flaws on each strip chart image. Human operators also annotated the exact location of each flaw in each related strip chart with a bounding box. The recorded data include both the raw inspected weld data and the interpreted weld data. From the weld inspection datasets, we collected 359 B-scan strip chart images. The distribution of our B-scan dataset covers Fill 1, Hot pass 1, and Hot pass 2 zones. Figure 9.3 represents the 1920 × 1080 image from one of the recorded weld data files.

Each 1920 × 1080 image contains 23 or 27 strip charts. These strip chats can display A-scan, B-scan, and TOFD ultrasonic images. In this study, our primary focus was on the B-scan images within the strip charts. Manually cropping each of these strip charts from the main image is time-consuming and prone to inaccuracy. To address this, we developed an automated Python script for the task. The script crops the strip charts from the main image automatically. We collected 359 B-scan images from the main images. We used DarkMark (`https://github.com/stephanecharette/DarkMark`, accessed on 15 August 2023, v1.8.11-1, Stéphane Charette, West Kelowna, BC, Canada) software to annotate flaws on the raw B-scan

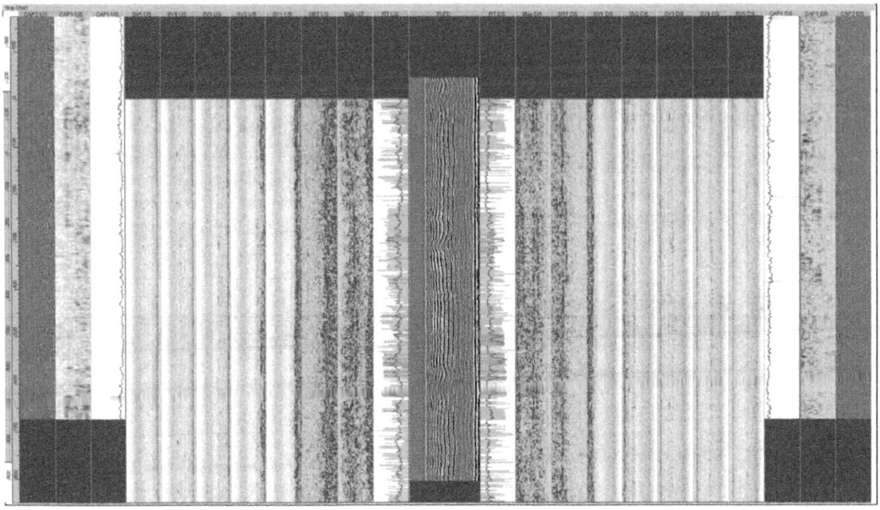

Figure 9.3 Ultrasonic data acquired from weld zones.

Table 9.1
Training, validation, test set image quantity.

Split	Images W Annot.	Images W/o Annot.	Total Images	Total Annot.
Training	95	243	338	182
Validation	10	0	10	16
Test	11	0	11	31

images based on human operator interpretations. For training,and evaluation of deep learning models, we divided the dataset into training, validation, and test sets. In Table 9.1, we provide details about the quantity of images, annotations, and negative samples in each set. Our dataset contains a total of 229 LOF annotations across B-scan images. Figure 9.4 presents 7 B-scan images and 10 LOF annotations from the training set.

In the final subsection, we analyze our B-scan dataset images to gain valuable insights. We believe that this information helped us with fine-tuning and interpreting the utilized deep-learning-based model more effectively.

9.2.3 DATASET PROPERTIES

The primary inspiration for this section of the study comes from `https://deci.ai/course/profiling-computer-vision-datasets-overview` (accessed on 17 November 2023). We only used their descriptions for image analysis to

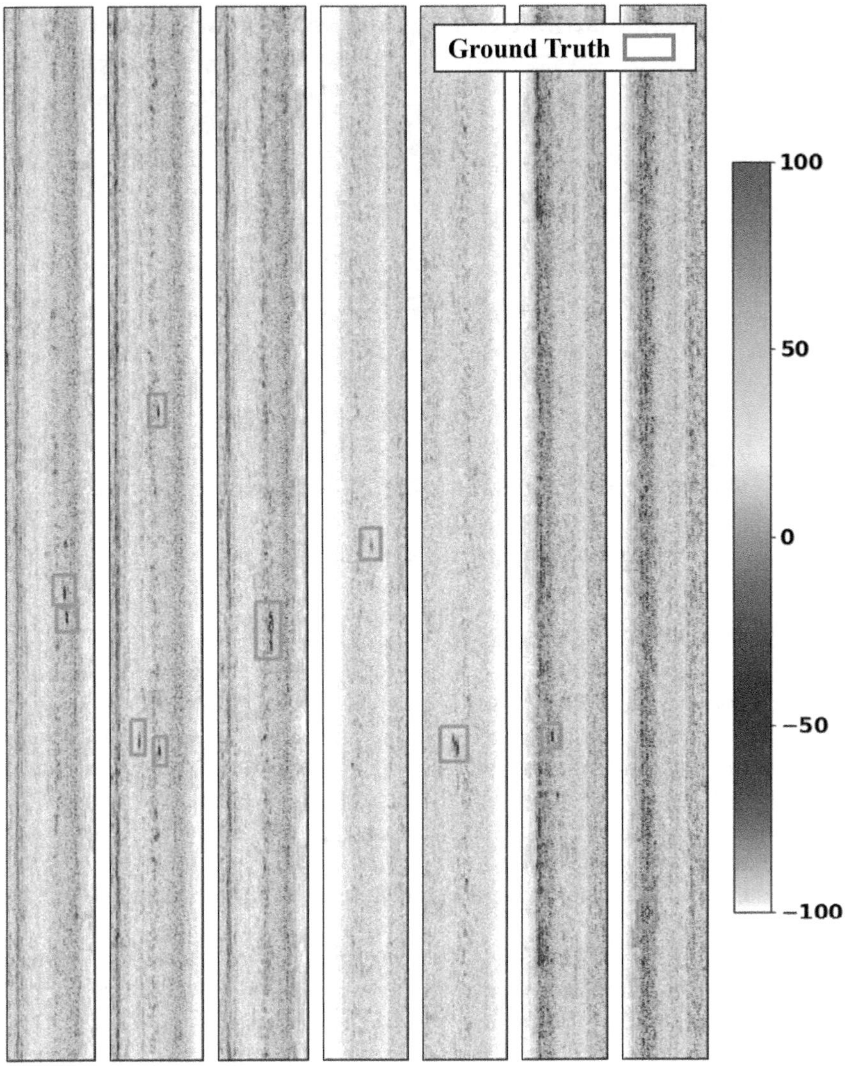

Figure 9.4 Samples of B-scan images from training set.

implement them based on our B-scan dataset. We did not use their Python package for generating our figures in this article.

9.2.3.1 Image Dataset Analysis

In this section of this study, our objective was to analyze the images themselves, disregarding the bounding boxes used for defect annotations. We analyzed these images based on their brightness distribution, width and height distribution, and mean and

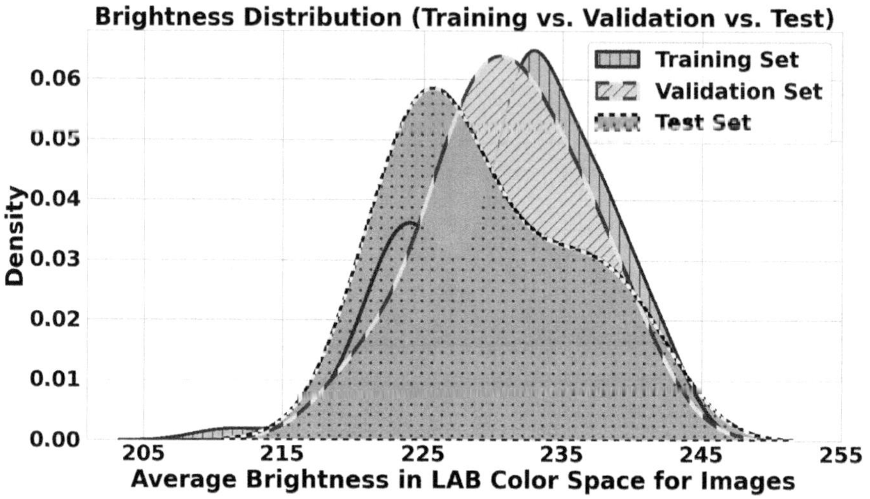

Figure 9.5 Brightness distribution in each set.

standard deviation (STD) of the pixels in the images. To begin, we focused on brightness distribution. For brightness analysis, we converted each image from the RGB to the LAB color space and calculated the average of the lightness (L) component. The resulting brightness distribution for all three sets is plotted in a kernel density (KDE) plot in Figure 9.5, revealing that the brightness distribution in the validation and test sets was mostly consistent with that of the training set images.

Next, we examined the dimensions distribution of all images. Figure 9.6 shows the number of images with specific width or height sizes for the three sets, demonstrating that the resolutions of images from the validation and test sets are included in the test set.

The other properties we calculated for our dataset were mean (Used "Mean pixels for normaliztion.ipynb" script from this repo `https://github.com/iDataVisualizationLab/roaddamagedetector` (accessed on 26 January 2024) for pixel mean calculations) and STD of pixels over three image channels including red, green, and blue. These properties could be helpful for more customization of the deep-learning-based object detector model, which was pretrained on the ImageNet [23] and COCO [24] dataset. Table 9.2 includes the mean and STD of pixels for the images in each set in our dataset.

In the next section, we analyze the bounding box features based on our dataset.

9.2.3.2 Bounding Box Analysis

For the analysis of the implemented bounding boxes for annotations, we discuss the drawing bounding boxes based on the area of each defect as a percentage of the total B-scan image, their counts in each image, and the spatial distribution of defects in

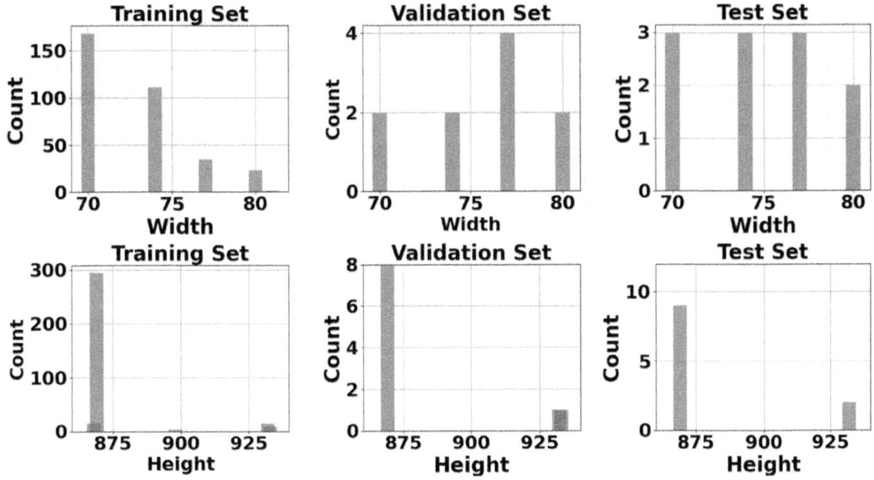

Figure 9.6 Image size distribution in each set.

Table 9.2
Mean and STD of pixels in our dataset in RGB format.

Split	Mean	STD
Training	[214.52, 230.39, 247.45]	[32.78, 23.22, 12.52]
Validation	[214.15, 230.02, 247.26]	[35.03, 24.86, 12.75]
Test	[211.56, 228.64, 246.99]	[33.21, 23.26, 12.21]

the images in our dataset. We performed these analyses for our training, validation, and test sets.

To analyze the distribution of bounding box area for each image in each set, the area of each bounding box was calculated as a percentage of the total B-scan image area. Figure 9.7 represents the LOF defects' area distribution in our dataset. Based on the plots, it is noticeable that we faced the challenge of detecting small defects in our B-scan image dataset. The minimum and maximum square roots of the bounding box area (anchor box size) in our training set were almost 16 and 32, respectively.

The bounding boxes' *height/width* ratio distribution (anchor box ratio) is represented in Figure 9.8. It indicates the anchor box ratio for training set was mainly focused in the middle of the distribution graph including 1.2, 1.4, and 1.7.

Moreover, we analyzed the bounding boxes based on their location in each image. In Figure 9.9, the defects are distributed throughout the images in each set. This can be advantageous for training a model with more generalization ability based on the consistently distributed defects in our dataset.

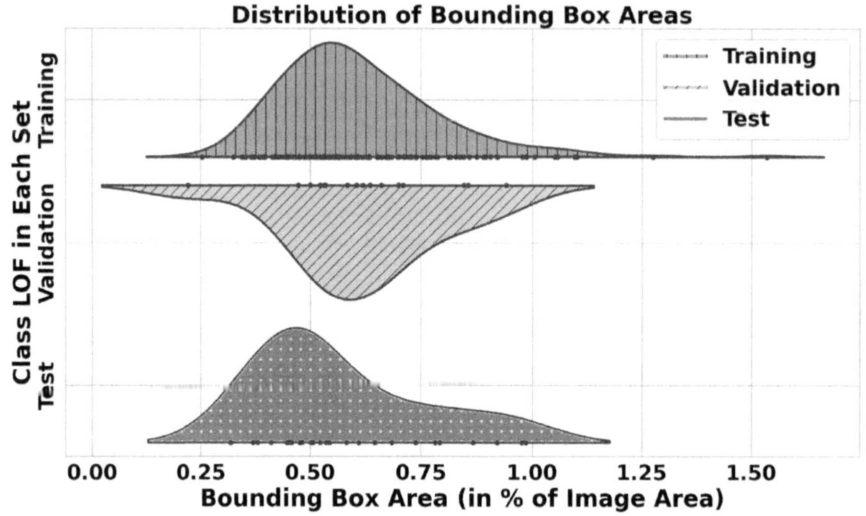

Figure 9.7 Bounding box area distribution.

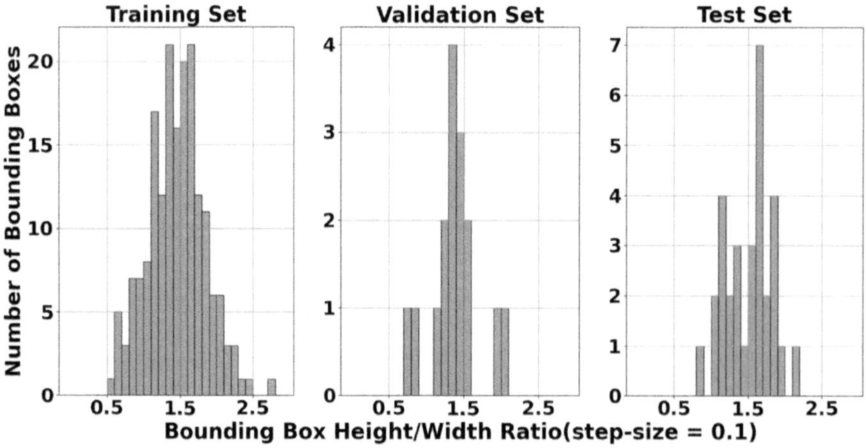

Figure 9.8 *Height/weight* ratio bounding box distribution.

Finally, Figure 9.10 demonstrates the number of bounding boxes appearing in each set of images. Through this analysis, we gained an idea of the maximum number of defects that a model should detect in an image based on our dataset.

9.3 EXPERIMENTAL RESULTS

In this section, we first introduce the models selected for our experiments. Then, we delve into the configurations and hyperparameters that were set manually for each

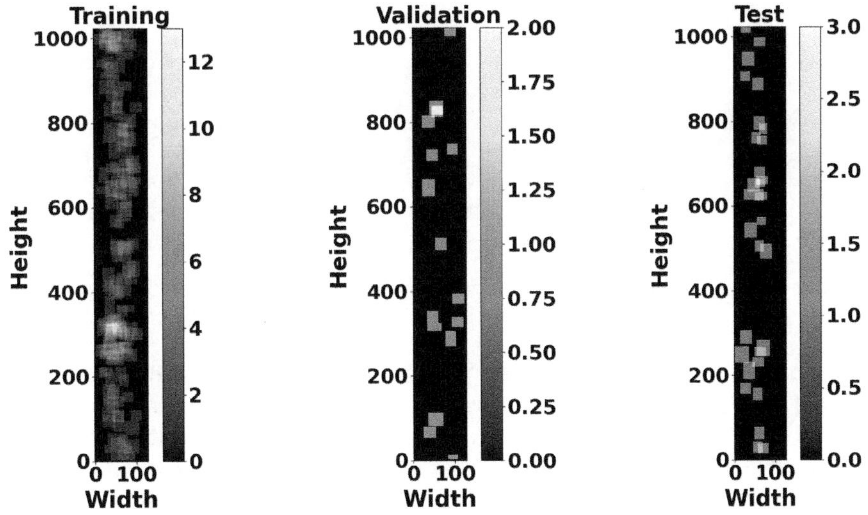

Figure 9.9 Spatial distribution of bounding boxes in the images in each set.

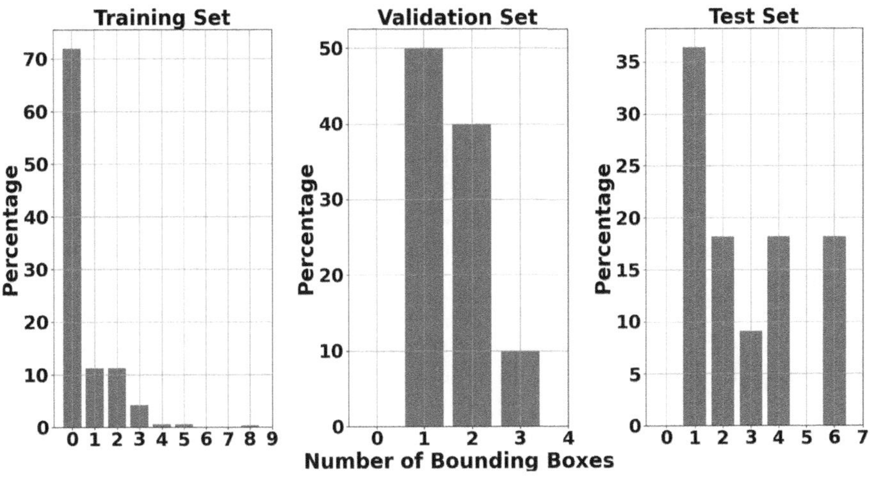

Figure 9.10 Distribution of available bounding boxes in each image in different sets.

model during training and evaluation. Finally, we compare the performance of these models on our validation and test sets.

We selected five SOTA object detectors to fine-tune them on our task. The models for our experiments included RetinaNet [25], DETR [26], Deformable DETR [27], YOLOv5u [28], and YOLOv8 [29]. These models mainly comprise CNN-based (RetinaNet, YOLOv5u, and YOLOv8) and transformer-based (DETR and Deformable DETR) object detectors. The reasoning behind choosing these models is

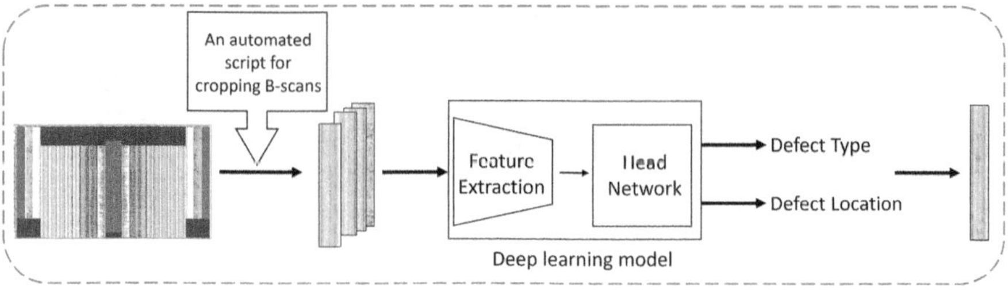

Figure 9.11 An overview of the methods applied to our dataset for detecting flaws.

that, first, we included RetinaNet and YOLOv5, which have been studied in previous works [16, 17] on B-scan images, to evaluate their performance on our dataset. Moreover, the performance of end-to-end transformer-based object detectors (DETR and Deformable DETR) and recent CNN-based architectures (YOLOv8) has not been previously investigated on B-scan images by the research community. Therefore, the aim of the selection of these diverse architectures was to broaden our understanding of the performance and impact of different types of deep-learning-based object detectors for the UT flaw detection task.

In the following, we elaborate on the configurations and hyperparameters that we manually set for each model during the training and evaluation stages. The main hyperparameters we manually set were the learning rate, input image size to the networks, and the numbers of iterations and epochs. The rationale behind choosing these was to set them based on the properties of our image dataset. Any configurations and hyperparameters not explicitly mentioned were kept consistent with the choices made by the model developers during the pretraining phase. Our objective in this step was to maintain most of the settings used for pretraining the models, aiming to establish the preliminary results of these models on our dataset as the baseline. All models were fine-tuned and evaluated on a single Nvidia A100 40 GB GPU with Cuda 11.8, torch 2.1.2, and torchvision 0.16.2 [30]. To enhance the repeatability of the results across different hyperparameters, we set the random seed to 42 for more deterministic random initializations. Notably, we chose not to train these models from scratch, as discussed in [31], to avoid unnecessarily prolonging the training process without significant improvements in model performance and due to the absence of a sufficiently large dataset for pretraining. Figure 9.11 represents the streamline through which B-scans passed to be analyzed for potential defects.

9.3.1 RETINANET

RetinaNet is a CNN-based single-stage detector that employs a novel loss function called focal loss to address the class imbalance problem during training. One of the key features of RetinaNet is its ability to detect objects across a wide range of scales by using feature pyramid networks [32] and anchors of multiple scales and aspect ratios. We utilized its implementation on Detectron2 [33] with an ImageNet-pretrained ResNet-50 [34] backbone. The initial configuration change involved enabling the model to train on negative samples and avoiding the filtering of empty samples (We set `_C.DATALOADER.FILTER_EMPTY_ANNOTATIONS` to `False` in `detectron2/config/defaults.py` and `filter_empty` argument of `get_detection_datasets_dicts` to `False` in `detectron2/data/build.py`). Flaws in B-scan images can be easily misinterpreted as background due to the complexity of the images, even by humans. By doing so, we allow the model to learn features from backgrounds with nonexistent flaws. We retained the default implemented image augmentation methods for the training, validation, and test sets. For the training set, scale augmentation was applied, resizing the input images so that the shortest side was at least 640 and at most 800 pixels, while the longest side was at most 1333 pixels. In addition, a random horizontal flip with a probability of 0.5 was applied to the training set images. For evaluation on the validation and test sets, images were resized to 800 × 1333 using scale augmentation. The total batch size was set to 64, and training utilized stochastic gradient descent (SGD) [35] with a learning rate of 1×10^{-4} and a weight decay of 0. The training ran for 5000 iterations. The learning rate scheduler increased the initial learning rate from 3×10^{-5} to 1×10^{-4} linearly in the first 100 iterations and then maintained the learning rate at 1×10^{-4} for the next 3200 iterations. Subsequently, the learning rate decayed by a factor of 0.1 in the next 1100 iterations. Finally, the training continued for the last 600 iterations with a learning rate of 1×10^{-6}. To prevent overfitting, we monitored both training and validation loss (with the help of this github repo: `https://github.com/ravijo/detectron2_tutorial` (accessed on 11 December 2023), we logged validation loss). The Detectron2 platform logged validation loss, applying the same implemented image augmentations for the training set for the validation set. Additionally, the best checkpoint hook with the AP50 metric was incorporated into the training script to track the best model with improved AP50 results on the validation set every 50 iterations. We selected the model with the highest AP50 performance at iteration 1450, before the model started overfitting.

9.3.2 END-TO-END TRANSFORMER-BASED OBJECT DETECTORS

For these two models, there are common tuned configurations, which we discuss before addressing the hyperparameters specific to each model separately in their respective sections.

We fine-tuned two transformer-based object detectors, including DETR and Deformable DETR, on our B-scan dataset using their implementations on Detrex [36] codebase based on Detectron2. This framework facilitates the comparison of the performance among different DETR-based models. Both models exploit the ImageNet-pretrained ResNet-50 as their backbone. We exclusively applied the same image augmentation technique to all three image sets, involving rescaling to 96 × 1152 through scale augmentation. Additionally, both models were trained on both negative and positive samples by disabling the filtering of empty samples. The total batch size was set to 64, and training was optimized with AdamW [37], setting the base learning rate to 1×10^{-5} (in the DETR paper [26], the base learning rate is referred to as the transformer's learning rate), B1 = 0.9, B2 = 0.999 with no weight decay, keeping the learning rate constant during training. Similar to RetinaNet on Detectron2, we incorporated the best check point hook to save the model with the highest AP50 performance on validation set every 50 iterations. We also monitored training and validation loss to prevent overfitting. For DETR, the training ran for 10,000 iterations, while for Deformable DETR, it was 2000 iterations. We did not use the gradient clipping and expotional moving average (ema) configurations available in the Detrex codebase in our experiments.

9.3.2.1 DETR

The DETR model is the first end-to-end object detector that uses a transformer architecture [38], departing from the traditional convolutional approach. It treats object detection as a direct set-to-set prediction problem, where a set of object queries attends to the encoded image and outputs a set of predictions in parallel. A key feature of DETR is its simplicity and scalability, as it does not use hand-crafted components like anchor boxes or nonmaximum suppression (NMS). We fine-tuned two different implementations of DETR, namely, DETR-R50 and DETR-DC5-R50. The former is DETR with ResNet-50, while the latter involves DETR with increased feature resolution, which is achieved by adding dilation to the last stage of the backbone and removing a stride from the first convolution of this stage (dilated C5 stage) [26]. The number of object queries was set to 100, which is the default value for DETR. Additionally, the learning rate for the backbone, following the default DETR implementation, was 1×10^{-5}. For DETR, the model with the highest AP50 was achieved at the 8100th iteration, and for DETR-DC5, the best AP50 was attained at iteration 5350.

9.3.2.2 Deformable-DETR

Deformable DETR is an extension of the DETR model, introducing deformable attention modules that enable attention to a small set of key sampling points around a reference point. This deformable attention mechanism enhances the performance on object detection tasks, particularly for objects with irregular shapes or occlusions, while maintaining the end-to-end simplicity of the original DETR architecture. We fine-tuned three different variants of Deformable DETR, including Deformable DETR-Res50 (DDR50), DDR50 with box refinement, and the two-stage DDR50 with box refinement. The box refinement is an iterative bounding box refinement aimed at improving detection performance. In the two-stage edition, region proposals are generated by a variant of Deformable DETR in the first stage. Then, by feeding the generated proposals to the decoder as object queries for further refinement, the two-stage Deformable DETR is formed [27]. For the number of object queries and backbone learning rate, we followed the Detrex implementations (in the original implementation [27], the number of object queries is 300, the base learning rate is 2×10^{-4}, and the learning rates for the backbone, reference points, and sampling offset are set to 2×10^{-5}. In the Detrex implementation, the backbone learning rate is 1×10^{-5}, while the learning rates for the sampling offsets and reference points are set to be the same as the base learning rate. For our experiment, the base learning rate was configured to 1.0×10^{-5}) by setting it to 300 and 1×10^{-5}, separately. We identified the best model based on the AP50 performance of the three variants of DDR50 at iterations 1600, 700, and 1100.

9.3.3 YOLO FAMILY

To fine-tune YOLOv5u and YOLOv8, we used their official implementation by Ultralytics. The only parameter we considered in our experiments was the patience parameter, which is for early stopping of training if no improvement is observed in the evaluation on the validation set based on the AP50 metric. We repeated each model's fine-tuning process with five patience levels [50, 100, 150, 180, 200, 250] and selected the fine-tuned model with the highest mAP50 on the validation set. We set the input image size parameter (imgsz) to 1024. Moreover, the training for both YOLO models and their variants in our experiments ran with the SGD optimizer (learning rate = 0.01, momentum = 0.9, and no weight decay). We did not add any predefined image augmentation techniques to the model, and the default implemented augmentation methods by Ultralytics used for the model, e.g., online imagespace and colorspace augmentations (discussed in `https://github.com/ultralytics/yolov5/discussions/10469`, accessed on 15 December 2023). In the following, we discuss which variants of each model we chose for our experiments.

9.3.3.1 YOLOv5u

In general, YOLOv5 builds upon previous versions of the CNN-based You Only Look Once (YOLO) [39] model. It employs several novel techniques such as mosaic data augmentation, self-adversarial training, and adaptive anchors to improve performance. We used the nano edition of YOLOV5u, which is the optimized implementation of the original YOLOV5, benefiting from an anchor-free detection head inspired by YOLOv8. In our experiments, we referred to them as YOLOv5nu and YOLOV5n6u. The former was pretrained on input images with a size of 640 × 640, and the latter with 1280 × 1280. The best weight for YOLOv5nu occurred after 308 epochs with a patience level of 180. For YOLOv5n6u, it occurred after 219 epochs with a patience level of 180.

9.3.3.2 YOLOv8

YOLOv8 features an anchor-free detection head and optimizes the accuracy–speed trade-off by introducing several novel convolutional-based modules based on the YOLOv5 architecture (For more details, please refer to `https://github.com/ultralytics/ultralytics/issues/189`, accessed on 14 May 2024) [29, 40]. To fine-tune, we used its nano and small variants, YOLOv8n and YOLOv8s. The former's best weight was captured after 196 epochs with a patience level 100. For the latter, it occurred after 139 epochs with a patience level of 50.

9.3.4 EVALUATION

All fine-tuned models were evaluated on our validation and test sets. Three parameters, including true positive (*TP*), false negative (*FN*), and false positive (*FP*), were calculated for the images in each set. The number of flaws available in the images that are correctly identified by a model is called true positive (*TP*). The flaws that the model fails to identify are false negatives (*FN*s). False positive (*FP*) occurs when the model mistakenly identifies a flaw that is not actually present in the image. Based on these parameters, accuracy, precision, recall, and F1 score were calculated to evaluate the performance of the models using

$$\text{F1 score} = 2 \cdot \frac{\text{Precision} \cdot \text{Recall}}{\text{Precision} + \text{Recall}} \tag{9.1}$$

where

$$\text{Accuracy} = \frac{TP}{TP+FP+FN}\text{ ; Precision} = \frac{TP}{TP+FP}\text{ ; Recall} = \frac{TP}{TP+FN} \tag{9.2}$$

The metrics we used for our evaluations were primarily those employed in previous works [9, 41, 42] for assessing deep-learning-based defect detectors on their respective datasets. Before the calculations of *TP*s, *FP*s, and *FN*s in the evaluation stage, we applied two postprocessing steps to the predictions of each model. First, we considered a threshold value for the confidence (score) property of each prediction. Predictions with a confidence less than the determined threshold value were ignored.

The thresholds were not equal across the models and were selected from a range of [0, 1.0] with a step of 0.05 based on each model's best performance in the accuracy metric (To calculate the *TP*s, *FP*s, and *FN*s with different confidence thresholds for YOLO families, we used the scripts from [43] (`https://github.com/Cartucho/mAP`, accessed on 10 January 2024). In Table 9.3, the column "C.T." under each image set represents the selected confidence threshold (C.T.) for that model, indicating the model showing the best performance based on accuracy among the observed thresholds. Additionally, a model prediction was considered a *TP* when the intersection of the ground truth box and the predicted box was at least a 0.5 intersection over the area of the union of the two mentioned boxes, which is called intersection over union (IOU). In the current stage of our work, after each model made its predictions, in cases of missed or incorrect flaw detections, the predictions were evaluated by UT human experts to finalize the results.

9.4 RESULTS AND DISCUSSION

To demonstrate the performance and learning capabilities of the fine-tuned models, we compare their performance on the training set and validation set with their loss plots in Figure 9.12. To ensure a fair comparison, the term 'loss' refers to the total loss for each model. For RetinaNet, the total loss includes class loss and bounding box regression loss; for DETR-based models, it includes class loss, bounding box loss, and GIoUs loss; for YOLO families, it includes class loss, bounding box loss,

Table 9.3
Comparison of the performance of the SOTA models on the detection of the lack of fusion (LOF) flaw on both validation and test sets. The top performance is highlighted in bold, and the second-best is indicated by underline.

		Validation Set								Test Set							
Model	**Params**	**C.T.**	***TP***	***FP***	***FN***	**ACC**	**P**	**R**	**F1**	**C.T.**	***TP***	***FP***	***FN***	**ACC**	**P**	**R**	**F1**
RetinaNet-Res50	34 M	0.35	14	5	2	0.667	0.737	0.875	0.800	0.30	21	17	10	0.438	0.553	0.677	0.609
DETR-Res50	41 M	0.80	12	4	4	0.600	0.750	0.750	0.750	0.80	18	8	13	0.462	0.692	0.581	0.632
DETR-DC5-Res50	41 M	0.75	10	7	6	0.435	0.588	0.625	0.606	0.70	17	9	14	0.425	0.654	0.548	0.596
Deformable DETR-Res50	40 M	0.35	12	3	4	0.632	0.800	0.750	0.774	0.45	17	3	14	0.500	0.850	0.548	0.667
+bbox refinement	40 M	0.35	13	3	3	0.684	0.813	0.813	0.813	0.20	16	10	15	0.390	0.615	0.516	0.561
++two-stage	40 M	0.10	15	5	1	0.714	0.750	**0.938**	0.833	0.15	21	7	10	0.553	0.750	0.677	0.712
YOLOv5nu	2.6 M	0.50	13	4	3	0.650	0.765	0.813	0.788	0.15	23	10	8	0.561	0.697	0.742	0.719
YOLOv5n6u	4.3 M	0.50	13	3	3	0.684	0.813	0.813	0.813	0.15	26	11	5	0.619	0.703	**0.839**	0.765
YOLOv8n	3.2 M	0.40	15	2	1	**0.833**	0.882	**0.938**	**0.909**	0.35	24	4	7	**0.686**	0.857	0.774	**0.814**
YOLOv8s	11.2 M	0.55	11	1	5	0.647	**0.917**	0.688	0.786	0.40	20	3	11	0.588	**0.870**	0.645	0.741

and directional feature learning (DFL) loss. Additionally, in Figure 9.13, we compare the training loss and AP50 performance for the fine-tuned models. Since RetinaNet and DETR-based models are iteration-based, while YOLO families are epoch-based, we compared them separately.

Based on the evaluation metrics discussed in the previous section, Table 9.3 presents the results of each fine-tuned model on both the validation and test sets. Among the evaluated models, YOLOv8n outperformed the others in terms of F1 score on the test set. This metric indicated a balanced performance in precision and recall, suggesting robust minimization of *FP*s and *FN*s. For the recall metric, YOLOv5n6u surpassed YOLOv8n, with 8.40% better performance. Recall measured how well a model avoided missing flaws. Regarding the number of *TP*s, YOLOv5n6u had the highest count on the test set. Moreover, in terms of precision, YOLOv8s outperformed YOLOv8n with only 1.52% better performance. Precision reflected the accuracy of the model in correctly identifying real flaws among all instances it predicted as flaws. YOLOv8n, which performed well in terms of F1 score and accuracy, belongs to the category of small-sized architectures. Additionally, the top three models with the lowest parameters exhibited better recall measurements than the others and became the models with the highest *TP* counts.

In Figure 9.14, the accuracy, precision–recall curve, and F1 score on both the validation and test sets for all the fine-tuned models are compared across confidence thresholds ranging from 0 to 1 with a step of 0.05. The four metrics were calculated at each confidence threshold where the model generated at least one correct prediction. In this figure, on the test set, YOLOv8s achieves the highest precision up to the first half of the recall values, while YOLOv8n leads in the second half. DETR-DC5 and DDR50 with the bbox refinement exhibit the lowest precision over recall. YOLOv5n6u outperforms the others up to a confidence threshold of 0.2 in F1 score, with YOLOv8n surpassing the others up to 0.75, and DETR leading for the rest. DETR-DC5 has the lowest F1 score for the first half of the confidence thresholds, while RetinaNet performs poorest for higher thresholds. The interpretations regarding F1 score are consistent with those of accuracy at different thresholds. In

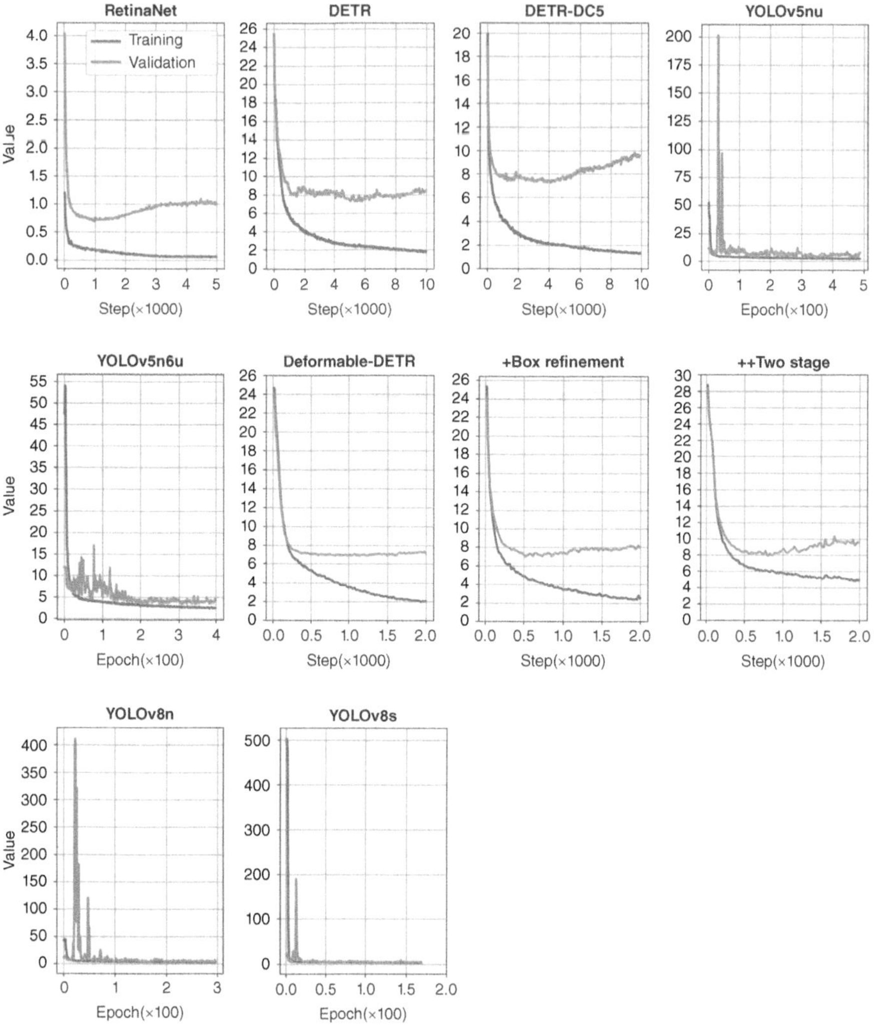

Figure 9.12 Comparison of training and validation losses for each model.

the results depicted in by Figure 9.15, the performance of the six models in detecting LOF flaws from a B-scan image on our test set is demonstrated.

We also investigated the effect of two parameters on the performance of RetinaNet and the DETR-based models. Initially, we repeated the fine-tuning and evaluation procedures by solely modifying the "mean_pixel" from its ImageNet mean_pixel to our train set image mean_pixel. Next, alongside the mean_pixel adjustment, we altered the "test_topK_candidates" (We modified the test_topK_candidates argument of RetinaNet class in detectron2/modeling/meta_arch/retinanet.py) from 1000 to 10 for RetinaNet and "num_queries" (We also set the select_box_nums_for_evaluation

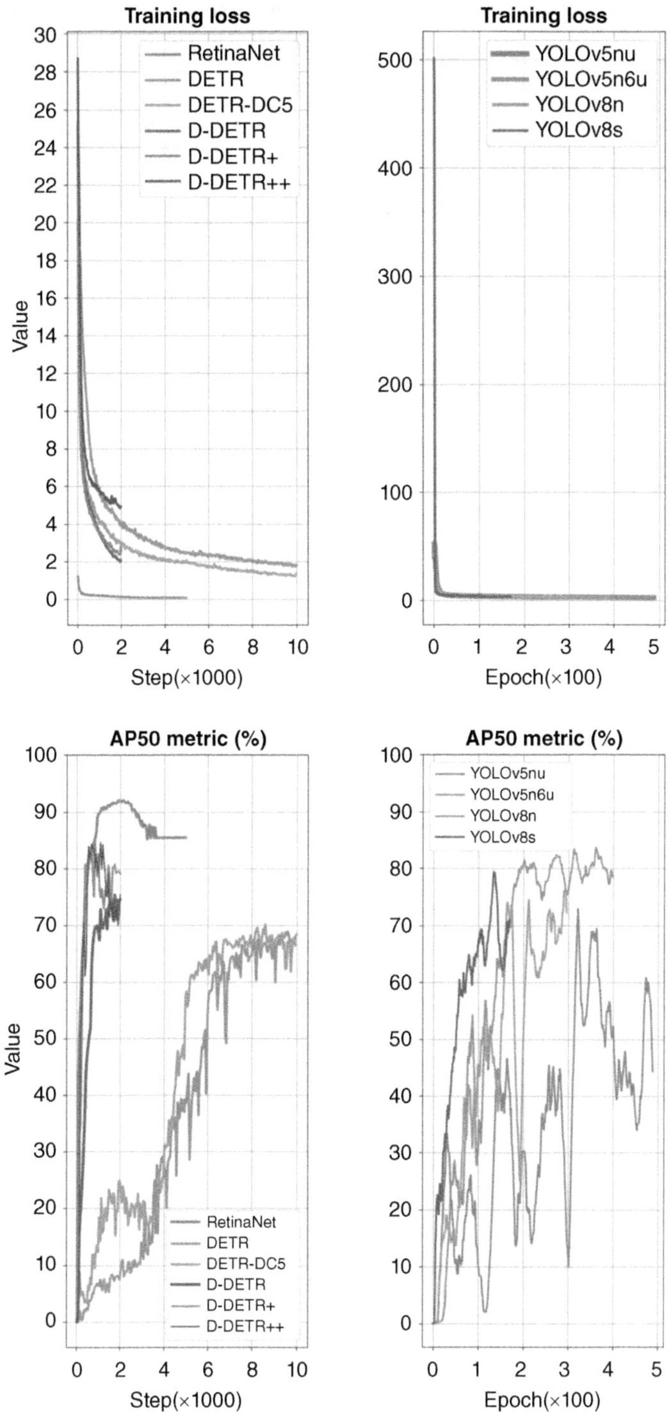

Figure 9.13 Comparing the training loss and AP50 metric performance during the training of all models.

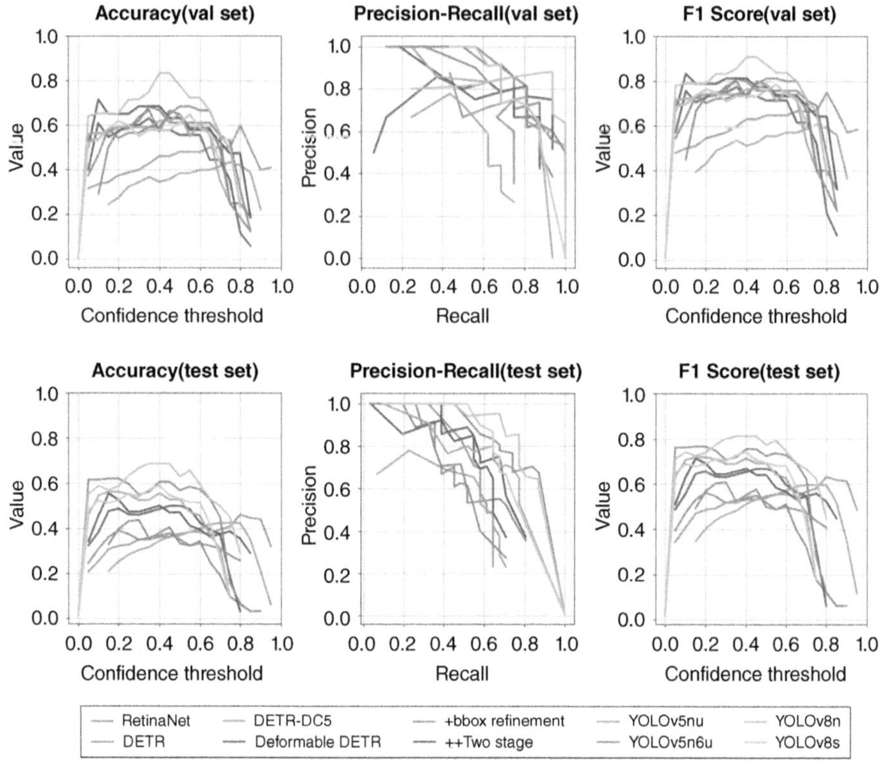

Figure 9.14 Comparing the accuracy vs. confidence thresholds, precision vs. recall, and F1 score vs. confidence thresholds on both validation and test sets for all models.

parameter of models to 10 to run the training and inference) from 100 to 10 for DETR-based models. In Table 9.4, we present the percentage of change relative to the reported values of accuracy, precision, recall, and F1 score in Table 9.3 on the test set. The results indicate that for DDR50 with box refinement, these modifications led to improvements in all four metrics. Furthermore, the impact of reducing "num_queries" in DETR-based models suggests the need to explore the effects of other parameters, such as "eos_conf", to enhance the results. This can be investigated in future studies.

In our B-scan images, discriminating an LOF flaw from weld geometry (non-flawed background image) was a challenging task, even for expert operators. This challenge can be addressed by providing more images containing both weld geometry indications and flaws during training. In our images, human operators did not annotate obscure indications of a flaw; however, some of the fine-tuned models detected these types of flaws, resulting in a high count of *FP*s. This understanding is grounded in the human knowledge and comprehension of the context of the image.

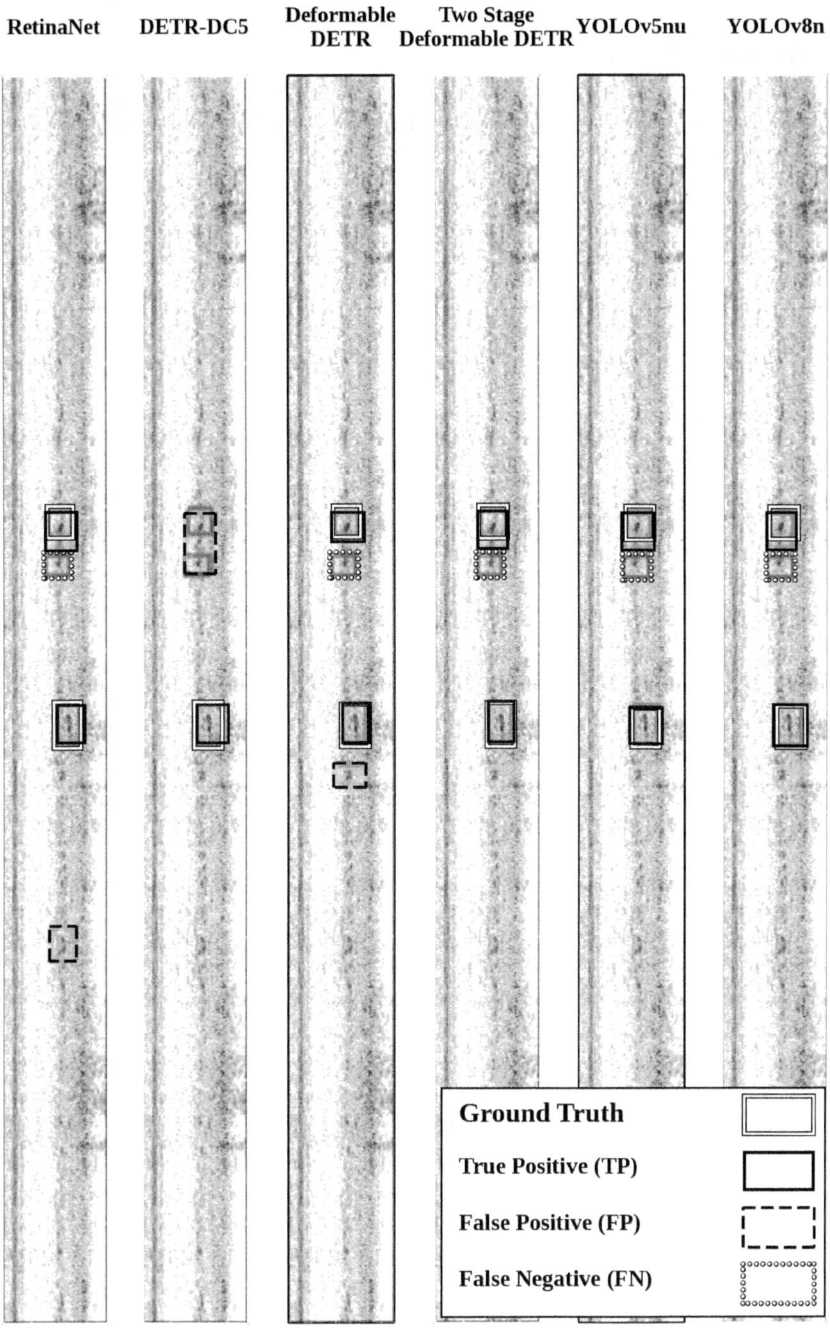

Figure 9.15 Performance comparison of SOTA object detectors on an image from test set. The B-scan contained 3 LOF flaws. The label on the top-left of each image represents the model name used for flaw prediction.

Table 9.4

The effect of modification of mean_pixel and topK or num_queries on the performance of the models on the test set. The bold text indicates an improvement relative to the value in Table 9.3 on the test set.

	mean_pixel				mean_pixel + (topK or num_queries)			
Model	ACC	P	R	F1	ACC	P	R	F1
RetinaNet-Res50	−13.75%	−13.83%	−4.76%	−9.98%	−12.61%	**+47.02%**	−38.10%	−9.12%
DETR-Res50	−16.67%	−5.80%	−16.67%	−12.04%	−45.83%	−33.80%	−38.89%	−36.67%
DETR-DC5-Res50	−21.57%	−14.35%	−17.65%	−16.18%	−32.03%	−26.36%	−23.53%	−24.85%
Deformable DETR-Res50	−21.21%	**+1.96%**	−23.53%	−15.22%	−53.85%	−37.72%	−47.06%	−43.75%
+bbox refinement	**+21.38%**	**+17.00%**	**+12.50%**	**+14.51%**	**+24.46%**	**+31.55%**	**+6.25%**	**+16.47%**
++two-stage	−11.97%	0.00%	−14.29%	−8.05%	−23.10%	−12.82%	−19.05%	−16.21%

Based on our evaluation of the performance of the fine-tuned models on industrial B-scan images, deploying these models as AI-assisted inspectors can enhance inspection speed and reduce the workload for human UT inspectors in industries utilizing AUT as one of their NDT methods. To address the challenges with industrial usage, one of the main considerations is deploying these deep-learning-based models on the hardware that runs inspection tools. Firstly, these models should be scaled and made more efficient in terms of inference speed and power usage to run on mobile devices, which are the most used in these industries. On the other hand, industries involved in designing NDT inspection tools should be open to providing open-source software that allows different industries to integrate their AI-based models with the acquisition software they use. In our implementation, due to the limitations of the acquisition software, we needed to develop additional software to access B-scan images and prepare them for feeding to the models, which would impact real-time inspection.

9.5 CONCLUSIONS

In this chapter, we presented a real-world ultrasonic B-scan image dataset, comprising 359 images and 229 annotations for lack of fusion (LOF) flaws. We conducted an extensive analysis of both the image properties and the bounding box properties utilized for annotations. To the best of our knowledge, this is the first time in the literature where industrial B-scan images containing real-world defects were employed for automated weld defect detection using SOTA deep learning models. End-to-end transformer-based object detectors and YOLOv8 were among the fine-tuned models evaluated on our dataset. The selected deep learning models were fine-tuned without any customization in their architectures to establish a baseline benchmark for these models on our dataset. Based on our evaluation, YOLOv8n exhibited superior performance compared to other fine-tuned models in terms of accuracy and F1 score. For the evaluations, we investigated the performance of two types of architectures: CNN-based and transformer-based architectures. According to our results, the CNN architectures with fewer model parameters, without any specific modifications, demonstrate superior performance, specifically compared to a few ten-million-parameter DETR-based models. Among the transformer-based architectures, two-stage Deformable DETR shows the best F1 score, which is 12.53% lower than that of YOLOv8n. Moreover, we demonstrated the effect of different confidence thresholds on the models' performance using accuracy, precision, recall, and F1 score metrics.

Limitations: While we demonstrated promising applications of deep learning models for industrial B-scan data acquired in the oil and gas industry, it is important to acknowledge the limitations of our work. Our dataset only covers one flaw type, LOF, hindering broader application. Additionally, the limited size of our dataset prevented us from providing a clear view of the inference time required by these models for real-time applications. Though we presented an AI-assisted inspection tool, human evaluation is necessary for missing or incorrect detections. To boost reliability, a human feedback loop may be needed for the model to learn from false positives and negatives. Finally, we did not investigate the effect of the model's hyperparameters and augmentation techniques on model performance, as we aimed to establish a baseline benchmark for these models. Therefore, we kept default options.

Future Works: For future studies, to enhance the diversity of our dataset, we plan to expand it by acquiring more industrial B-scan images collected during onshore automated girth weld inspections conducted by human UT inspectors. To improve the explainablity of the deep-learning-based models in this field, exploring the application of each layer and component of the models concerning how the B-scan images are processed through them, implementing automated hyperparameter tuning, and investigating the impact of various augmentation techniques on B-scan images, both during training and inference, are crucial for industrial applications. Additionally, studying the applications of new emerging AI-based models, including foundational models, in the field of AUT and analyzing B-scan data in conjunction with other types of ultrasonic data represent interesting ideas for enhancing automated defect detection.

REFERENCES

1. Cantero-Chinchilla, S.; Croxford, A.J.; Wilcox, P.D. Optimising laser-induced phased-arrays for defect detection in continuous inspections. *NDT E Int.* **2024**, *144*, 103091.
2. Aminzadeh, A.; Sattarpanah Karganroudi, S.; Barka, N.; El Ouafi, A. A real-time 3D scanning of aluminum 5052-H32 laser welded blanks; geometrical and welding characterization. *Mater. Lett.* **2021**, *296*, 129883. https://doi.org/10.1016/j.matlet.2021.129883.
3. Dwivedi, S.K.; Vishwakarma, M.; Soni, A. Advances and researches on non destructive testing: A review. *Mater. Today: Proc.* **2018**, *5*, 3690–3698.
4. Swornowski, P.J. Scanning of the internal structure part with laser ultrasonic in aviation industry. *Scanning* **2011**, *33*, 378–385. https://doi.org/10.1002/sca.20260.
5. Krizhevsky, A.; Sutskever, I.; Hinton, G.E. ImageNet Classification with Deep Convolutional Neural Networks. In Proceedings of the Advances in Neural Information Processing Systems, Lake Tahoe, CA, USA, 3–6 December 2012; Volume 25.
6. Ye, J.; Ito, S.; Toyama, N. Computerized Ultrasonic Imaging Inspection: From Shallow to Deep Learning. *Sensors* **2018**, *18*, 3820. https://doi.org/10.3390/s18113820.

7. Yashiro, S.; Takatsubo, J.; Miyauchi, H.; Toyama, N. A novel technique for visualizing ultrasonic waves in general solid media by pulsed laser scan. *NDT E Int.* **2008**, *41*, 137–144. https://doi.org/10.1016/j.ndteint.2007.08.002.
8. Yashiro, S.; Toyama, N.; Takatsubo, J.; Shiraishi, T. Laser-Generation Based Imaging of Ultrasonic Wave Propagation on Welded Steel Plates and Its Application to Defect Detection. *Mater. Trans.* **2010**, *51*, 2069–2075. https://doi.org/10.2320/matertrans.M2010204.
9. Ye, J.; Toyama, N. Benchmarking Deep Learning Models for Automatic Ultrasonic Imaging Inspection. *IEEE Access* **2021**, *9*, 36986–36994. https://doi.org/10.1109/ACCESS.2021.3062860.
10. Ye, J.; Toyama, N. USimgAIST. A Public Dataset for Research on Computer-Vision-Based Pattern Investigation for Ultrasonic Wave Propagation Images in Non- Destructive Testing (NDT). accessed 2023
11. Zhang, X.; Saniie, J. Reinforcement Learning Based Neural Architecture Search for Flaw Detection in Intelligent Ultrasonic Imaging NDE System. In Proceedings of the 2022 IEEE International Ultrasonics Symposium (IUS), Venice, Italy, 10–13 October 2022; pp. 1–4. https://doi.org/10.1109/IUS54386.2022.9957772.
12. Posilović, L.; Medak, D.; Subašić, M.; Petković, T.; Budimir, M.; Lončarić, S. Flaw Detection from Ultrasonic Images using YOLO and SSD. In Proceedings of the 2019 11th International Symposium on Image and Signal Processing and Analysis (ISPA), Dubrovnik, Croatia, 23–25 September 2019; pp. 163–168. https://doi.org/10.1109/ISPA.2019.8868929.
13. Redmon, J.; Farhadi, A. YOLOv3: An Incremental Improvement. *arXiv* **2018**, arXiv:1804.02767.
14. Liu, W.; Anguelov, D.; Erhan, D.; Szegedy, C.; Reed, S.; Fu, C.Y.; Berg, A.C. SSD: Single Shot MultiBox Detector. In Proceedings of the Computer Vision ECCV 2016, Amsterdam, The Netherlands, 11–14 October 2016; pp. 21–37.
15. Tan, M.; Pang, R.; Le, Q.V. EfficientDet: Scalable and Efficient Object Detection. In Proceedings of the IEEE/CVF Conference on Computer Vision and Pattern Recognition, Seattle, WA, USA, 14–19 June 2020.
16. Medak, D.; Posilović, L.; Subašić, M.; Budimir, M.; Lončarić, S. Automated Defect Detection From Ultrasonic Images Using Deep Learning. *IEEE Trans. Ultrason. Ferroelectr. Freq. Control* **2021**, *68*, 3126–3134. https://doi.org/10.1109/TUFFC.2021.3081750.
17. Medak, D.; Posilović, L.; Subašić, M.; Budimir, M.; Lončarić, S. DefectDet: A deep learning architecture for detection of defects with extreme aspect ratios in ultrasonic images. *Neurocomputing* **2022**, *473*, 107–115. https://doi.org/10.1016/j.neucom.2021.12.008.
18. Virkkunen, I.; Koskinen, T.; Jessen-Juhler, O.; Rinta-aho, J. Augmented Ultrasonic Data for Machine Learning. *J. Nondestruct. Eval.* **2021**,
emph40, 4.
19. Svahn, P.; Virkkunen, I.; Zettervall, T.; Snögren, D. The use of virtual flaws to increase flexibility of qualification. In Proceedings of the 12th European Conference on Non-Destructive Testing (ECNDT 2018), Gothenburg, Sweden, 11–15 June 2018; Volume 23.
20. Posilović, L.; Medak, D.; Subašić, M.; Budimir, M.; Lončarić, S. Generating ultrasonic images indistinguishable from real images using Generative Adversarial Networks. *Ultrasonics* **2022**, *119*, 106610. https://doi.org/10.1016/j.ultras.2021.106610.

21. Posilović, L.; Medak, D.; Subašić, M.; Budimir, M.; Lončarić, S. Generative adversarial network with object detector discriminator for enhanced defect detection on ultrasonic B-scans. *Neurocomputing* **2021**, *459*, 361–369. https://doi.org/10.1016/j.neucom.2021.06.094.
22. Park, T.; Liu, M.Y.; Wang, T.C.; Zhu, J.Y. Semantic Image Synthesis with Spatially-Adaptive Normalization. In Proceedings of the IEEE/CVF Conference on Computer Vision and Pattern Recognition (CVPR) 2019, Long Beach, CA, USA, 15–20 June 2019.
23. Deng, J.; Dong, W.; Socher, R.; Li, L.J.; Li, K.; Fei-Fei, L. Imagenet: A large-scale hierarchical image database. In Proceedings of the 2009 IEEE Conference on Computer Vision and Pattern Recognition, Miami, FL, USA, 20–25 June 2009; pp. 248–255.
24. Lin, T.; Maire, M.; Belongie, S.J.; Bourdev, L.D.; Girshick, R.B.; Hays, J.; Perona, P.; Ramanan, D.; Dollár, P.; Zitnick, C.L. Microsoft COCO: Common Objects in Context. *arXiv* **2014**, arXiv:1405.0312.
25. Lin, T.Y.; Goyal, P.; Girshick, R.; He, K.; Dollár, P. Focal Loss for Dense Object Detection. *arXiv* **2017**, arXiv:1708.02002.
26. Carion, N.; Massa, F.; Synnaeve, G.; Usunier, N.; Kirillov, A.; Zagoruyko, S. End-to-End Object Detection with Transformers. *arXiv* **2020**, arXiv:2005.12872.
27. Zhu, X.; Su, W.; Lu, L.; Li, B.; Wang, X.; Dai, J. Deformable DETR: Deformable Transformers for End-to-End Object Detection.
28. Jocher, G. ultralytics/yolov5: v3.1 - Bug Fixes and Performance Improvements. *J. Zenodo* **2020** https://doi.org/10.5281/zenodo.4154370. DOI: 10.5281/zenodo.4154370 (accessed on 27 November 2023)
29. Jocher, G.; Chaurasia, A.; Qiu, J. Ultralytics YOLOv8. *J. Comput. Commun.* **2023**, *11*. (accessed on 27 November 2023)
30. Paszke, A.; Gross, S.; Massa, F.; Lerer, A.; Bradbury, J.; Chanan, G.; Killeen, T.; Lin, Z.; Gimelshein, N.; Antiga, L.; et al. PyTorch: An Imperative Style, High-Performance Deep Learning Library. In Proceedings of the Advances in Neural Information Processing Systems 32, Vancouver, BC, Canada, 8–14 December 2019; pp. 8024–8035.
31. He, K.; Girshick, R.; Dollar, P. Rethinking ImageNet Pre-Training. In Proceedings of the IEEE/CVF International Conference on Computer Vision (ICCV), Seoul, Republic of Korea, 27 October–2 November 2019.
32. Lin, T.Y.; Dollár, P.; Girshick, R.; He, K.; Hariharan, B.; Belongie, S. Feature pyramid networks for object detection. In Proceedings of the IEEE Conference on Computer Vision and Pattern Recognition, Honolulu, HI, USA, 21–26 July 2017; pp. 2117–2125.
33. Wu, Y.; Kirillov, A.; Massa, F.; Lo, W.Y.; Girshick, R. Detectron2. 2019. Available online: https://github.com/facebookresearch/detectron2 (accessed on 11 October 2023).
34. He, K.; Zhang, X.; Ren, S.; Sun, J. Deep residual learning for image recognition. In Proceedings of the IEEE Conference on Computer Vision and Pattern Recognition, Las Vegas, NV, USA, 27–30 June 2016; pp. 770–778.
35. Ruder, S. An overview of gradient descent optimization algorithms. *arXiv* **2016**, arXiv:1609.04747.
36. Ren, T.; Liu, S.; Li, F.; Zhang, H.; Zeng, A.; Yang, J.; Liao, X.; Jia, D.; Li, H.; Cao, H.; et al. detrex: Benchmarking Detection Transformers. *arXiv* **2023**, arXiv:2306.07265.
37. Loshchilov, I.; Hutter, F. Fixing Weight Decay Regularization in Adam. *arXiv* **2017**, arXiv:1711.05101.

38. Vaswani, A.; Shazeer, N.; Parmar, N.; Uszkoreit, J.; Jones, L.; Gomez, A.N.; Kaiser, Ł.; Polosukhin, I. Attention is all you need. In Proceedings of the Advances in Neural Information Processing Systems 30 (NIPS 2017), Long Beach, CA, USA, 4–9 December 2017; Volume 30.
39. Redmon, J.; Divvala, S.; Girshick, R.; Farhadi, A. You only look once: Unified, real-time object detection. In Proceedings of the IEEE Conference on Computer Vision and Pattern Recognition, Las Vegas, NV, USA, 27–30 June 2016; pp. 779–788.
40. Contributors, M. MMYOLO: OpenMMLab YOLO Series Toolbox and Benchmark. 2022. Available online: `https://github.com/open-mmlab/mmyolo` (accessed on 15 May 2024).
41. Naddaf-Sh, S.; Naddaf-Sh, M.; Zargarzadeh, H.; Dalton, M.; Ramezani, S.; Elpers, G.; Baburao, V.S.; Kashani, A.R.; et al. Real-time explainable multiclass object detection for quality assessment in 2-dimensional radiography images. *Complexity* **2022**, *2022*, 4637939.
42. Lindgren, E.; Zach, C. Industrial X-ray Image Analysis with Deep Neural Networks Robust to Unexpected Input Data. *Metals* **2022**, *12*, 1963. `https://doi.org/10.3390/met12111963`.
43. Cartucho, J.; Ventura, R.; Veloso, M. Robust Object Recognition Through Symbiotic Deep Learning In Mobile Robots. In Proceedings of the 2018 IEEE/RSJ International Conference on Intelligent Robots and Systems (IROS), Madrid, Spain, 1–5 October 2018; pp. 2336–2341.

10 SAM for Weld Defect Detection in Ultrasonic B-Scans

Amir-M. Naddaf-Sh, Vinay S Baburao, Hassan Zargarzadeh

CHAPTER SUMMARY

Automated ultrasonic testing (AUT) is a critical tool for infrastructure evaluation in industries such as oil and gas, and, while skilled operators manually analyze complex AUT data, artificial intelligence (AI)-based methods show promise for automating interpretation. However, improving the reliability and effectiveness of these methods remains a significant challenge. This study employs the Segment Anything Model (SAM), a vision foundation model, to design an AI-assisted tool for weld defect detection in real-world ultrasonic B-scan images. It utilizes a proprietary dataset of B-scan images generated from AUT data collected during automated girth weld inspections of oil and gas pipelines, detecting a specific defect type: lack of fusion (LOF). The implementation includes integrating knowledge from the B-scan image context into the natural image-based SAM 1 and SAM 2 through a fully automated, promptable process. As part of designing a practical AI-assistant tool, the experiments involve applying both vanilla and low-rank adaptation (LoRA) fine-tuning techniques to the image encoder and mask decoder of different variants of both models, while keeping the prompt encoder unchanged. The results demonstrate that the utilized method achieves improved performance compared to a previous study on the same dataset. The code is available at: https://github.com/amirmohammadnsh/SAM-AUT.

10.1 INTRODUCTION

Non-destructive testing (NDT) techniques are non-intrusive methods for inspecting components and structures, ensuring their integrity and quality without disrupting daily operations in crucial systems. NDT is widely used in safety-critical industries, such as aerospace, automotive, and oil and gas, where structural reliability is vital. NDT encompasses various analysis methods, including ultrasonic testing, X-radiography, visual inspection, eddy current testing, magnetic particle testing, etc. NDT inspections generate vast amounts of data, which are typically interpreted manually by human operators, a process that is time-consuming, expensive, and prone to limitations due to the repetitive evaluations and high-dimensional data involved.

DOI: 10.1201/9781003766575-10

The advent of machine learning (ML) algorithms, particularly deep learning [1], enables automated data interpretation methods to achieve human-level performance in NDT, presenting a promising solution to this challenge. This is evident in applications of deep learning methods, such as automated weld defect detection in digital X-radiography images [2, 3, 4, 5], automated defect detection in ultrasonic testing [6, 7, 8, 9], and automated data evaluation in visual inspection [10, 11].

Ultrasonic testing (UT) is a widely used NDT technique, offering numerous benefits, including requiring access to only one side of the object being inspected, providing internal images of structures [12], and enabling precise subsurface small defect localization [13, 14, 9]. To further optimize ultrasonic testing inspections, researchers are investigating cutting-edge UT-based techniques aimed at increasing confidence and efficiency. These include optimizing laser-induced phased arrays (LIPAs) configurations for continuous inspections, striking a balance between detectability of target defects and acquisition time [15]. Additionally, a directivity calculation framework based on the reciprocity theorem has been developed, which is crucial for designing non-destructive evaluation inspections utilizing emerging technologies like laser ultrasound (LU) [12, 16]. Researchers are also investigating techniques to suppress surface acoustic waves (SAW) and enhance the signal-to-crosstalk ratio when using LIPAs for inspecting additive manufacturing processes [17].

In UT, a surface-mounted transducer sends ultrasound waves through a material, allowing the detection and evaluation of internal flaws and corrosion caused by manufacturing, welding, or other processes. Automated ultrasonic testing (AUT) employs automated systems with phased array probes, performing rapid and precise inspections. Each element in the array is pulsed and delayed independently, creating adjustable beam angles and focal points for comprehensive material examination. In ultrasonic testing (UT), data are typically captured as A-scans, B-scans, and C-scans. Inspired by AlexNet's [18] success in computer vision, researchers have leveraged deep learning algorithms to improve efficiency and reliability in downstream tasks, particularly in interpreting ultrasonic B-scan images, which provide more detailed visualizations of inspected materials.

The application of deep learning-based object detectors with convolutional neural network (CNN) architecture on ultrasonic B-scan images was first explored in [19]. DefectDet [20], a novel architecture, replaces the default backbone of EfficientDet [21] with a lightweight encoder–decoder-based feature extractor and a custom detection head for detecting defects with extreme aspect ratios in B-scan images. To address data scarcity, an extensive data augmentation method based on virtual flaws [22] is utilized in phased-array ultrasonic data to train a CNN-based architecture for defect classification. A more advanced solution involves generating realistic synthetic images using Generative Adversarial Networks (GANs) [23], which were applied for the first time to B-scan images in [24, 25]. The results showed that combining real and synthetic data can improve the performance of a deep learning-based object detector. Self-supervised learning methods were first applied to ultrasonic image datasets with limited positive samples in [26], using custom implementations of Ganomaly [27], PaDiM [28], and DifferNet [29] to detect defects in B-scan images, with models trained solely on normal images. Similarly, [30] employed a modified variational autoencoder (VAE) [31] to detect anomalies based on deviations in VAE outputs for defective images. To interpret a sequence of B-scan images simultaneously, two novel architectures based on merging high-dimensional features extracted from B-scan images with EfficientDet, using either Conv2D or ConvLSTM [32], are proposed in [33], addressing the fact that some defects are more apparent at specific angles or in an image sequence. In [34], the authors implemented a customized spatio-temporal CNN model for interpreting ultrasonic wave propagation image series, trained on their 1700 annotated clips.

Despite advancements in deep learning-based weld defect detection methods, their effectiveness often depends on customized architectures to achieve state-of-the-art results using synthetic data and artificially generated flaws in controlled laboratory settings. This dependency poses significant challenges for deploying these methods to assist human inspectors. In previous work [35] by the current authors, in-house real-world ultrasonic B-scan images from onshore inspections of oil and gas pipelines, containing a specific type of genuine flaw, Lack of Fusion (LOF), were utilized for the first time in the literature to fine-tune existing deep learning-based object detection methods without any customization. F1-scores of 0.712 and 0.814 were achieved with Two-stage Deformable DETR [36] and YOLOv8-nano [37], respectively. This chapter demonstrates that SAM 1 with the ViT-Base image encoder and SAM 2 with the Hiera-Base+ can achieve an F1-score of nearly 0.940 on the same dataset, while easing implementation difficulties compared to previously studied methods.

With the advent of deep learning techniques, foundation models [38], such as GPT-4 [39], Llama 3 [40], Gemini 1.5 [41], DALL-E [42], SegGPT [43], etc., have demonstrated promising performance in tasks related to natural language and image processing, thanks to their exceptional generalization capabilities and the vast amounts of data on which they were pre-trained. The Segment Anything Model (SAM) series, which includes SAM 1 [44] and SAM 2 [45], are vision foundation models designed for promptable visual segmentation in images and videos. These models return segmentation maps based on given prompts (e.g., points, boxes, masks). SAM 1 is trained on 11 million images and 1.1 billion masks (SA-1B dataset [44]), and SAM 2 is pre-trained on 51,000 diverse videos and 643,000 masklets (SA-V dataset [45]). This extensive training enables excellent generalization and zero-shot knowledge transfer capabilities, making them promising for downstream tasks, particularly in industrial applications. Many recent studies have demonstrated the applicability of the SAM series in various practical scenarios, including interpreting clinical medical images [46, 47, 48, 49], remote sensing [50], anomaly detection in industrial images [51], defect segmentation in thermal images [52], and civil infrastructure defect assessment [53].

In this study, the SAM family is utilized to design a reliable AI-assisted tool for automated weld defect detection in real-world ultrasonic B-scan images, where obtaining defective samples is nearly impossible due to procedures being designed to be defect-free. The dataset from a previous work [35], which includes only B-scan images of oil and gas pipelines with a single type of defect, is used. Building on similar work, such as SamLP [54], which adapted SAM 1 for license plate detection, this approach extends the methodology to handle B-scan images containing multiple defects of the same class within a single image. Notably, the implementation eliminates the need for morphological operations in post-processing by directly converting segmented regions into bounding boxes, yielding a robust solution for real-world applications. The key contributions of this chapter can be summarized as follows:

- This chapter proposes a fully automated and promptable defect detection approach using SAM 1 and SAM 2, while minimizing post-processing steps and parameters compared to existing deep learning-based methods (e.g., tuning confidence thresholds, applying non-maximum suppression (NMS), etc.).
- It investigates the effects of different fine-tuning methods (vanilla and low-rank adaptation (LoRA)), various scales of training data (to address limited data scenarios), and the impact of combining loss functions (Dice loss and cross-entropy).

In the experiments, stratified five-fold cross-validation is employed to fine-tune the image encoder and mask decoder of four different variants of each model, while keeping the prompt encoder frozen. The approach is illustrated in Figure 10.1, where SAM is fine-tuned on the ultrasonic B-scan dataset. During the post-processing stage, defects in the B-scan images are identified using bounding boxes generated from the mask decoder's segmentation outputs. Finally, human operators review and validate the model's results to address any missed or incorrect detections—a critical step, as the primary goal is to assist human operators in their daily tasks.

The remainder of this chapter is organized as follows: Section 10.2 provides details on the method, SAM 1's and SAM 2's architecture, and the fine-tuning methods utilized. Section 10.3 explains the implementation of the experiments and hyperparameter configuration. Section 10.4 presents the performance comparison of different variants of SAM 1 and SAM 2 on the B-scan dataset, discusses limitations, and outlines future studies. Finally, Section 10.5 concludes this chapter.

10.2 METHOD

10.2.1 OVERVIEW

Given a B-scan image $I_{B-Scan} \in \mathbb{R}^{H \times W \times C}$ (a 2D RGB image, explained in detail in Section 10.3.1), with spatial resolution $H \times W$ and C channels, the goal is to predict its corresponding segmentation map $\hat{S}$ (with resolution $H \times W$), where each pixel belongs to one of the predefined classes in $L = \{B, D\}$, specifically background (B) or defect (D). The prediction aims to closely match the ground truth S, using a bounding box prompt that indicates the entire picture in the format $[X_{top-left}, Y_{top-left}, X_{bottom-right}, Y_{bottom-right}]$, which, in this case, would be $[0, 0, H, W]$. As SAM 1 and SAM 2 lack knowledge of the context of ultrasonic B-scan images, as well as the shape and location of defects, fine-tuning becomes

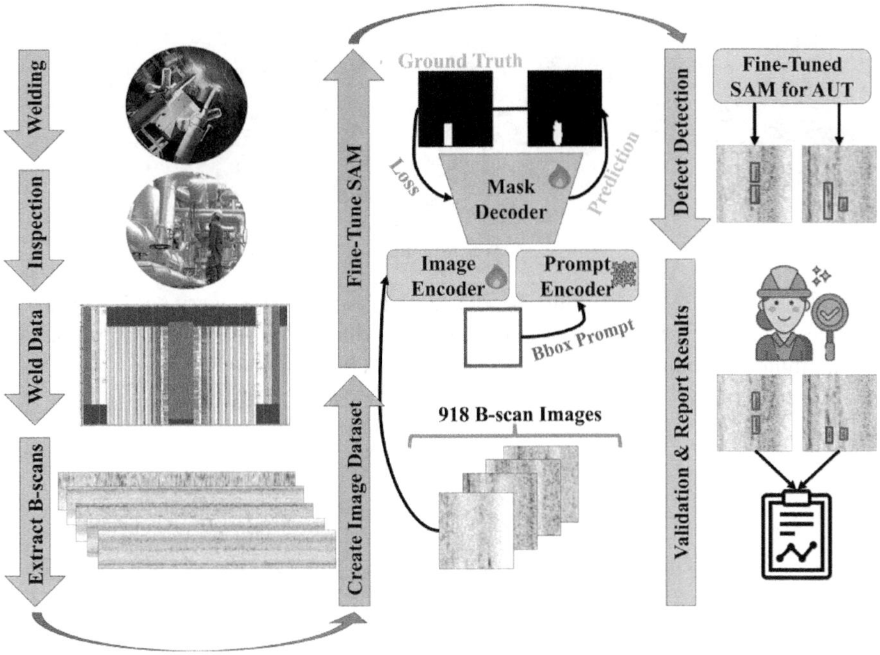

Figure 10.1 Overview of the study.

a necessary step. These two models output three segmentation masks to resolve segmentation prompt ambiguities. In this study, only the first predicted segmentation logit, with a of size $h \times w$ is utilized. This logit is aligned with the original input size using bi-linear interpolations, and the regions indicated as defects in each mask, are converted into bounding boxes for defect detection purposes. In the following sections, the architectures of SAM 1 and SAM 2, along with the fine-tuning strategies are discussed in more detail.

10.2.2 ARCHITECTURE

As shown in Figure 10.2, SAMs consist of three main components: an image encoder, a prompt encoder, and a mask decoder. SAM 2 enhances the SAM 1 architecture by incorporating memory capabilities for video applications, utilizing memory attention, a memory encoder, and memory bank mechanisms. This study primarily focuses on fine-tuning the image encoder and mask decoder of each SAM. Therefore, a detailed description of these components is provided, along with high-level information about the prompt encoder and the remaining blocks.

10.2.2.1 Image Encoder

For SAM 1's image encoder, both the vision transformer (ViT) [55] and TinyViT [56] backbones are fine-tuned. The main image encoder of SAM 1 uses a ViT pre-trained

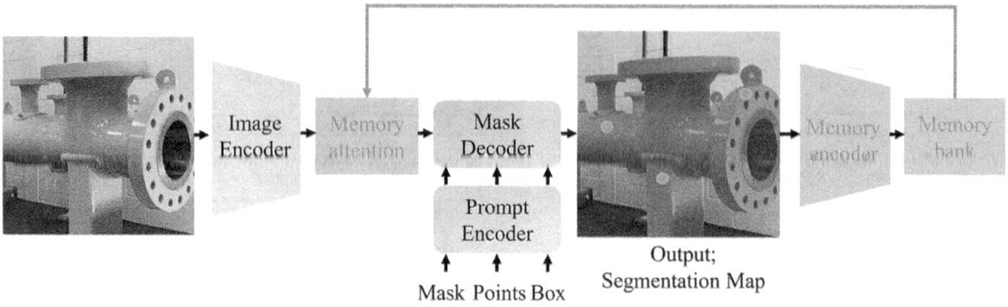

Figure 10.2 Segment Anything Model (SAM) architecture overview. The grayed-out blocks and arrow are bypassed for image handling. The yellow pipe is segmented using multiple point prompts (indicated by green dots).

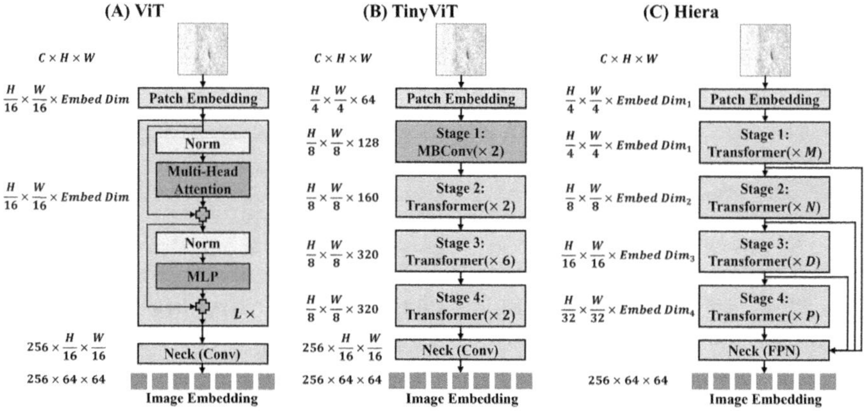

Figure 10.3 (**A**) Overview of image encoder of SAM 1 with ViT backbone. (**B**) Overview of image encoder of SAM 1 with TinyViT backbone. (**C**) Overview of image encoder of SAM 2 with Hiera backbone.

with Masked Autoencoders (MAE) [57]. As shown in Figure 10.3A, the input image passes through a patch embedding block, transformer blocks, and a neck block. The patch embedding block uses a trainable 16 × 16 convolution with a distinct embedding dimension (*Embed Dim*) for each ViT variant. This block generates image patch embeddings for the transformer blocks. Adapted from [58], these transformer blocks are basic transformer encoders [59], customized for high-resolution images using 14 × 14 windowed attention and four equally spaced global attention blocks. The neck block then uses 1 × 1 and 3 × 3 convolutions with 256 channels, each followed by layer normalization [60], to reduce the channel dimension. The resulting image encoder output is a 16× downscaled embedding, yielding a 64 × 64 representation.

Table 10.1
Configurations for ViT variants, TinyViT, and Hiera variants.

Model	#Channels	#Blocks	#Heads	Global Attn. Blocks
ViT-Base	768	12	12	[2-5-8-11]
ViT-Large	1024	24	16	[5-11-17-23]
ViT-Huge	1280	32	16	[7-15-23-31]
TinyViT	[128-160-320]	[2-6-2]	[4-5-10]	—
Hiera-T	[96-192-384-768]	[1-2-7-2]	[1-2-4-8]	[5-7-9]
Hiera-S	[96-192-384-768]	[1-2-11-2]	[1-2-4-8]	[7-10-13]
Hiera-B+	[112-224-448-896]	[2-3-16-3]	[2-4-8-16]	[12-16-20]
Hiera-L	[144-288-576-1152]	[2-6-36-4]	[2-4-8-16]	[23-33-43]

Table 10.1 provides detailed specifications for each ViT variant, including the number of transformer blocks, channels, and multi-head attentions.

The SAM 1 image encoder with a TinyViT backbone, introduced in [61], distills knowledge from ViT-H into TinyViT with 5M parameters. As shown in Figure 10.3B, TinyViT has a hierarchical structure with four main stages. Downsampling using MBConvs [62] occurs after each stage, except the last. Before Stage 1, the patch embedding block contains two 3×3 convolutions; the first is followed by a GELU [63] activation layer, and each is followed by BatchNorm [64]. In Stage 1, two MBConvs with an embedding dimension of 64 capture low-level features. Table 10.1 details configurations for subsequent transformer stages. The window sizes in each transformer stage are 7×7, 14×14, and 7×7, to reduce computation. The neck block is identical to that used in SAM 1's original image encoder.

SAM 2's image encoder adopts a hierarchical structure, leveraging the Hiera architecture [65], which builds upon MViTv2 [66], a ViT-based hierarchical model comprising four stages. Notably, Hiera employs convolution-free, pure hierarchical ViT blocks trained with MAE [57], omitting shifted, cross-shaped windows and decomposed relative position embeddings. Efficiency is achieved through local attention within "mask units" in the first two stages and global attention for the rest blocks. At each stage transition, features from Q and skip connections are doubled via linear layers, while spatial dimensions undergo 2×2 max pooling. Taking inspiration from [67], SAM 2 incorporates interpolated global positional embeddings for spatial information. The patch embedding structure resembles the patch embedding used in SAM 1's image encoder with ViT. The neck utilizes a Feature Pyramid Network (FPN) [68] to fuse features from Stages 3 (stride 16) and 4 (stride 32) of the Hiera encoder, generating image embeddings. Stride 4 and 8 features from Stages 1 and 2 enhance the mask decoder's upsampling layers. Table 10.1 provides detailed specifications for Hiera variants, including channel counts (*Embed Dim*), transformer blocks and multi-head attentions.

10.2.2.2 Prompt Encoder and Mask Decoder

There are two sets of prompts for SAMs: sparse (points, boxes) and dense (masks). Points (e.g., positive or negative clicks) and boxes are represented by positional encodings [69] summed with learned embeddings for each prompt type. Masks are embedded using convolutions and summed element-wise with the image embedding. The architecture of the prompt encoder is identical in both SAMs. Before running the mask decoder, a learned output token embedding is inserted into the prompt embeddings to be used at the decoder's output, similar to the [*class*] token in ViT, which serve as "tokens" (not including image embeddings).

In SAMs, as shown in Figure 10.4, the lightweight mask decoder combines image embeddings and tokens to produce an output mask. It utilizes a modified two-layer transformer decoder [59], where each layer performs four steps: self-attention on the tokens, cross-attention from tokens to image embeddings, updating each token using a point-wise multi-layer perceptron (MLP), and cross-attention from image embeddings to tokens. The subsequent decoder layer processes the updated tokens and image embeddings from the previous layer. After decoding, the updated image embedding is upscaled by $4\times$ using two transposed convolutional layers, and the tokens interact with it once more. The updated output token embedding is then passed through a three-layer MLP, whose output is combined with the upscaled image embedding via a point-wise product, yielding the final predicted mask.

For SAM 2's mask decoder, there are two skip connections from stride 4 and 8 features from Stages 1 and 2 of the image encoder to the upscaling layers in the mask decoder, incorporating high resolution information for mask decoding. Additionally, two modifications are applied to the mask decoder to make it more robust for video applications. First, an occlusion prediction head is introduced by applying an MLP to the new token, the occlusion token, to indicate the likelihood of the object being visible in the current frame. Second, an object token pointer is placed in the memory bank for each frame.

10.2.2.3 Memory Attention, Memory Encoder, and Memory Bank

Regarding SAM 2's memory capability, the memory encoder stores past prediction mask memories in the memory bank. Subsequently, memory attention combines these stored memories with the current frame's embedding, generated by the image encoder, producing a conditioned embedding that is then passed to the mask decoder.

10.2.3 VANILLA AND PARAMETER-EFFICIENT FINE-TUNING

SAMs are pre-trained on natural images; therefore, adopting them for downstream tasks requires transferring domain knowledge through fine-tuning methods. This transfer learning is performed using the vanilla fine-tuning [70] and parameter-efficient fine-tuning (PEFT) methods. In vanilla fine-tuning, all image encoder and

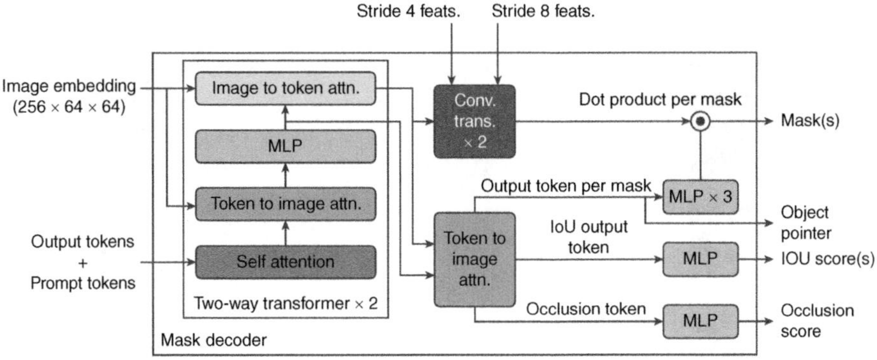

Figure 10.4 Overview of the mask decoder architecture in SAMs.

mask decoder parameters are updated, while the parameters of the remaining blocks in each SAM are kept frozen.

Parameter-efficient fine-tuning (PEFT) [71] methods offer an alternative to vanilla fine-tuning, reducing computational costs and preventing overfitting and catastrophic forgetting [72]. This approach involves freezing most parameters while selectively fine-tuning a subset of the original model's parameters. Low-rank adaptation (LoRA), a popular PEFT method is employed to fine-tune SAMs. Inspired from low "intrinsic dimension" in pre-trained language models [73], LoRA hypothesizes low "intrinsic rank" for weight updates and adds low-rank matrices to the self-attention layers in transformer blocks [74, 75, 76]. As shown in Equation (10.1), LoRA freezes the pre-trained weight matrix $W_0 \in \mathbb{R}^{d\times k}$ and decomposes the parameter update matrix ΔW into two trainable low-rank matrices, $B \in \mathbb{R}^{d\times r}$ and $A \in \mathbb{R}^{r\times k}$, where matrix B is zero, A is initialized with random Gaussian distribution in the beginning, and the rank $r \ll min(d,k)$. The input feature $x \in \mathbb{R}^{n\times d}$ is multiplied with both W_0 and $\Delta W = BA$, and their outputs are summed element-wise to produce $h \in \mathbb{R}^{n\times k}$.

$$h = W_0 x + \Delta W x = W_0 x + BAx \tag{10.1}$$

10.3 EXPERIMENTS

10.3.1 DATASET

In this chapter, the dataset of ultrasonic B-scan images, introduced in a previous study [35], is utilized. The B-scan images were collected from proprietary weld inspection records generated by UT experts using the inspection system during onshore oil and gas pipeline automated girth weld inspections with phased array technology. The inspection system employs the zone discrimination technique for weld inspection. In the dataset, the weld type is J-bevel, and the weld defect class is lack of fusion (LOF), which is the most common defect in automated girth welding. The

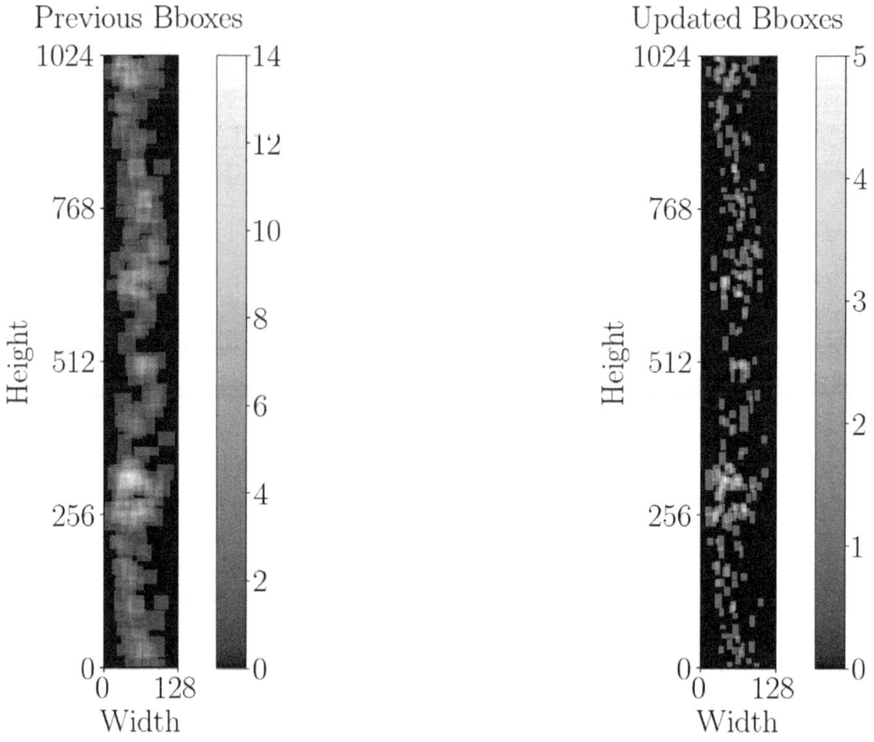

Figure 10.5 Comparing spatial distribution of bounding boxes in B-scan images with at least one defect, both before and after modification.

B-scan images primarily cover the body and part of the root regions of the weld (for further details, please refer to Section 2 of [35]).

For this study, the previous bounding box annotations are tightened using Labelme (5.4.1) [77] to minimize background inclusion. This modification ensures that, when converting bounding boxes to segmentation labels, the resulting labels accurately identify more pixels related to defects rather than background. Figure 10.5 illustrates the spatial distribution of bounding boxes in the B-scan images with at least one defect, both before and after modification. Subsequently, only the B-scan images containing at least one annotation, totaling 116 images with 229 annotations, are selected for this study. Then, 64×64 patches with 32 pixels overlap are created and zero-padding is applied to patches with less than 64 pixels to ensure uniform image dimensions. The 50% overlap can be interpreted as data augmentation for defects moving in consecutive patches. The resulting patches are filtered to include only those with at least one annotation in the final dataset, yielding in a total of 918 images with 1010 annotations.

To split data into training, validation, and test sets for the experiments, following the routine in ML-based studies, 20% of the dataset is allocated to the test set and

Table 10.2
#Images (#Annotations) by Set and Fold.

Split	Fold 1	Fold 2	Fold 3	Fold 4	Fold 5
Train	582 (645)	586 (645)	590 (646)	591 (646)	587 (646)
Val	152 (162)	148 (162)	144 (161)	143 (161)	147 (161)
Test			184 (203)		

the remaining portion are reserved for training and validation in a stratified manner. For the remaining data, stratified five-fold cross-validation is applied during the fine-tuning stage. In Table 10.2, the number of images and annotations for the training set and validation set of each fold, as well as the test set, are represented. The ratio of train, validation, and test sets in all folds is almost 4:1:1.25.

10.3.2 ENVIRONMENT

All models are fine-tuned using an Nvidia A100 40 GB GPU with CUDA 11.8 (cuDNN 8.9.7) on a machine running Ubuntu 20.04 LTS. The code is implemented in Python (3.10), using PyTorch (2.2.2) and Torchvision (0.17.2) [78] for SAM 1 (fine-tuned with FP32) and PyTorch (2.4.0) and Torchvision (0.19.0) for SAM 2 (fine-tuned with BFloat16 and automatic mixed precision (AMP)).

10.3.3 INPUT DATA PIPELINE

To meet the input requirements of SAM 1 and SAM 2, the same pre-processing steps introduced in each of these models are applied in this study. SAM 1 first rescales the longest side of all the RGB-based images to 1024 using bi-linear interpolation. Then, it normalizes the resized images by subtracting ImageNet [79] mean pixel values from image pixels in each channel and dividing them by the ImageNet standard deviation (STD) pixel values. Afterward, it applies zero-padding to the shortest side. However, since the B-scan images are initially $3 \times 64 \times 64$, rescaling yielded $3 \times 1024 \times 1024$ images without requiring padding. It then reorders the dimensions to $1024 \times 1024 \times 3$ to match its image encoder input ($H \times W \times C$).

To use SAM 1 for automated mask generation for the entire image, all bounding box prompts for each image are formatted as $[0, 0, W, H]$. After rescaling, this format becomes $[0, 0, 1024, 1024]$. Additionally, to match the output dimension of the SAM mask decoder, all available segmentation labels for each annotation in each image are combined, as the images may contain one or multiple annotations. The same transformation applied to the images is then applied to the labels, resizing them to 256×256.

For SAM 2, both sides of the images are resized to 1024 using bi-linear interpolation, and the pixel values are normalized using the ImageNet mean and STD pixel

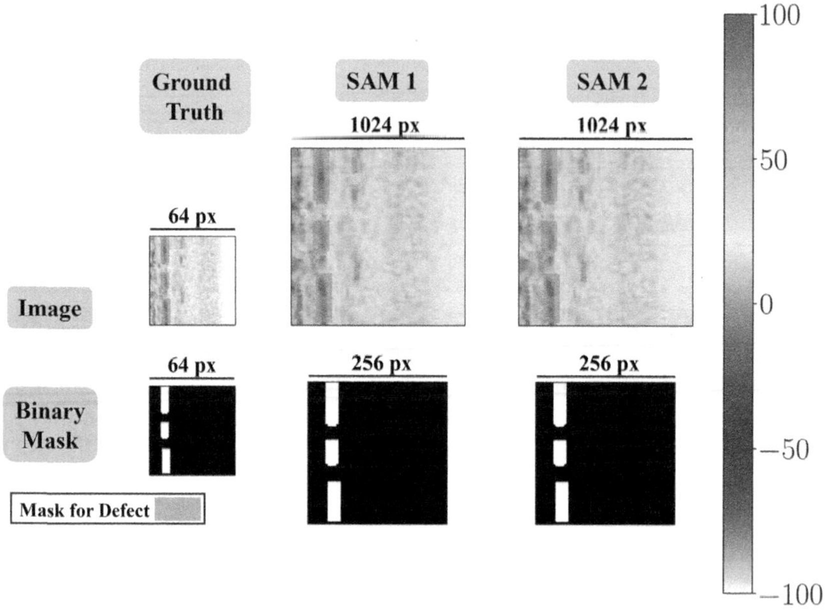

Figure 10.6 A 64 × 64 B-scan image and its corresponding mask after applying image and mask transformations for each SAM.

values. Similar to SAM 1, the bounding box prompts indicate the whole image, but in SAM 2 are in the format $[[0,0],[W,H]]$. Moreover, the segmentation labels are resized to 256 in both width and height using bi-linear interpolation. Figure 10.6 shows a sample of the final B-scan image and its mask, resulting from the steps described above, and used for fine-tuning each of the SAMs.

The SAM mask decoder outputs an embedding with a resolution of 256 × 256 (in single output mode). To match the original image size, this embedding undergoes bi-linear interpolations ($256 \times 256 \rightarrow 1024 \times 1024 \rightarrow 64 \times 64$), followed by sigmoid activation and a binary operation with a threshold of 0.5, resulting in a binary mask output. To obtain the bounding box coordinates of predicted defects, post-processing is conducted as follows. First, small disconnected regions and holes are removed from the mask using the "remove_small_regions" function from the SAM published code, with a minimum area threshold of 4.5. This threshold corresponds to the minimum defect area in B-scan images before patch conversion and almost twice the minimum defect area in 64 × 64 patches. Next, the refined masks are converted to bounding box using the "find_contours" method from the scikit-image library (0.24.0) [80]. The resulting bounding box, indicating a defect, is formatted as $[X_{top-left}, Y_{top-left}, X_{bottom-right}, Y_{bottom-right}]$.

10.3.4 EVALUATION METRICS

The performance of the fine-tuned SAMs on the dataset is evaluated using the common evaluation metrics in object detection tasks, including F1-score and average precision (AP). F1-score, as illustrated in Equation (10.2), considers both precision (P) and recall (R) simultaneously. Three parameters—true positive (TP), false negative (FN), and false positive (FP)—are used to calculate P and R. The defects available in the images that are correctly identified by a model are referred to as TPs. Defects that the model fails to identify are called FNs. An FP occurs when the model mistakenly identifies a defect that is not actually present in the image. Additionally, a model prediction was considered a TP when the intersection of the ground truth box and the predicted box was at least a 0.5 intersection over the union (IoU) of the two boxes (10.4).

$$\text{F1-score} = 2 \cdot \frac{P \cdot R}{P + R} \tag{10.2}$$

where

$$Acc = \frac{TP}{TP+FP+FN}\ ; P = \frac{TP}{TP+FP}\ ; R = \frac{TP}{TP+FN} \tag{10.3}$$

$$\text{IoU}(\text{bbox}_{\text{pred}}, \text{bbox}_{\text{gt}}) = \frac{\text{bbox}_{\text{pred}} \cap \text{bbox}_{\text{gt}}}{\text{bbox}_{\text{pred}} \cup \text{bbox}_{\text{gt}}} \tag{10.4}$$

The AP metric (10.5) averages precision values across recall rates. It calculates the area under the precision–recall (PR) curve by summing the products of $P(k)$ (the precision at the k-th recall rate, $R(k)$) and $\Delta R(k)$ (the difference between the k-th and ($k-1$)-th recall rates) [81].

$$\text{AP} = \sum_{k=1}^{n} P(k)\Delta R(k) \tag{10.5}$$

The F1-score and average precision (AP) are calculated by modifying the evaluation metrics implemented in the SamLP [54] scripts to align with the specific requirements of this study.

10.3.5 FINE-TUNING

10.3.5.1 Loss Function

The loss between the ground truth mask $G = \{g_i\} \in \mathbb{R}^{256\times256}$ and the predicted mask $S = \{s_i\} \in \mathbb{R}^{256\times256}$ is computed using Dice loss [82], a widely adopted and robust loss function for segmentation tasks [83]. s_i and g_i denoting the corresponding segmentation results of each pixel i in the N-pixel image. In general, Dice loss [84] is computed as

$$L_{\text{Dice}} = 1 - 2\frac{\sum_{i=1}^{N}\sum_{c=1}^{C} g_i^c s_i^c}{\sum_{i=1}^{N}\sum_{c=1}^{C}(g_i^c)^2 + \sum_{i=1}^{N}\sum_{c=1}^{C}(s_i^c)^2} \tag{10.6}$$

where C denotes the number of classes. For this chapter implementation, which focuses on a single class of defects, the equation simplifies to

$$L_{\text{Dice}} = 1 - 2\frac{\sum_{i=1}^{N} g_i s_i}{\sum_{i=1}^{N}(g_i)^2 + \sum_{i=1}^{N}(s_i)^2} \tag{10.7}$$

In this study, MONAI's (1.3.2) [85] Dice loss implementation with the sigmoid function is applied to the predictions.

10.3.5.2 LoRA Settings

LoRA is applied on the image encoder of SAM 1 using the SAMed implementation [76, 86, 75]. This implementation inserts LoRA layers into the "query" and "value" projection matrices within self-attention layers of specific transformer blocks. For the experiments in this study, the blocks featuring global attention are targeted. The rank value, r, is 4 across all experiments. For SAM 2's image encoder, LoRA implementation from the PEFT library (v0.13.0) [87] is used, applying it to the "query", "value", and "key" matrices in self-attention layers of transformer blocks with global attention, also with a rank of 4.

10.3.5.3 Fine-Tuning Overview

As discussed in Section 10.2.3, both vanilla and LoRA fine-tuning methods are utilized in the experiments. For vanilla fine-tuning, the configurations include SAM 1 with TinyViT (MobileSAM [61] checkpoint), and ViT-B, and SAM 2 with Hiera-T, Hiera-S, and Hiera-B+. To mitigate overfitting, LoRA fine-tuning is applied solely to the image encoder block in SAM 1 (ViT-L and ViT-H) and SAM 2 (Hiera-L) configurations, while their corresponding mask decoders are fine-tuned using the vanilla approach. The fine-tuning scripts are built upon the published code of recent studies that fine-tuned SAM 1 and SAM 2 for medical domains and other downstream tasks [46, 88, 54], incorporating required customizations for this study.

In the fine-tuning process for each model the ReduceLROnPlateau scheduler is used in 'min' mode with zero patience, adjusting the learning rate after each epoch if no loss reduction occurred. Optimization is performed using the AdamW optimizer ($B_1 = 0.9, B_2 = 0.999$ and weight decay 0.01). Moreover, each model is evaluated at the end of every epoch on the validation set during the fine-tuning process. The checkpoint achieving the highest F1-score and lowest loss on the validation set across all previous epochs is saved as the best model in each fold. The final best fine-tuned model is selected from the fold with the highest F1-score on the validation set, and this model is then evaluated on the test set.

To set the initial learning rate and learning rate decay factor of the scheduler for each model, initial learning rates of 10^{-5}, 10^{-4}, and 10^{-3} are tested, along with learning rate decay factors ranging from 0.6 to 0.9 (incrementing by 0.1). First, the fine-tuning process is initiated with a learning rate decay factor of 0.9, testing each learning rate. Then, the remaining decay factors are tested, using the corresponding learning rate that yielded the best F1-score on the test set. Furthermore, the decay

Table 10.3
Hyperparameters for fine-tuning.

Model	Image Encoder	Fine-Tune Mode	Epochs	Initial LR	LR Decay Factor	Batch Size
SAM 1	TinyViT	Vanilla	30	10^{-3}	0.7	16
	ViT-Base	Vanilla	25	10^{-4}	0.7	4
	ViT-Large	LoRA	15	10^{-3}	0.9	2
	ViT-Huge	LoRA	10	10^{-3}	0.9	1
SAM 2	Hiera-T	Vanilla	30	10^{-5}	0.8	8
	Hiera-S	Vanilla	30	10^{-5}	0.9	8
	Hiera-B+	Vanilla	25	10^{-5}	0.8	8
	Hiera-L	LoRA	10	10^{-3}	0.9	4

factor was decremented only if an improvement in the F1-score was observed on the test set. The learning rate configurations, along with the corresponding number of epochs and batch size, are presented in Table 10.3.

10.4 RESULTS AND DISCUSSION

10.4.1 PERFORMANCE EVALUATION

In Table 10.4, the results of the experiments are presented. Among SAM 1 variants, the ViT-B backbone achieves the best performance based on the F1-score, outperforming ViT-L, ViT-H, and TinyViT by 2.5%, 3.5%, and 5.3%, respectively. The lower performance of SAM 1 with the TinyViT backbone arises from using knowledge distillation from a larger teacher model; however, this is offset by a significant 89.2% reduction in the model's parameter count, which contributes to a decrease in memory usage. For SAM 1 with ViT-L and ViT-H backbones, the applied LoRA technique reduces the trainable parameters of the image encoders to around 100K while still benefiting from the pre-trained knowledge of these two backbones.

For the SAM 2 variants, the one with the Hiera-B+ backbone delivers the highest performance, exceeding the Hiera-T and Hiera-S variants by 1.4% and 7.43%, respectively. The SAM 2 variant with the Hiera-L backbone further reduces the number of trainable parameters by 85.50% through the use of the LoRA technique, but it lags 9.7% behind the variant with the Hiera-B+ backbone.

SAM 1 with the ViT-B backbone slightly outperforms SAM 2 with the Hiera-B+ backbone by 0.11% in F1-score. However, SAM 2 shows notable efficiency benefits, reducing training time by 80% and increasing inference speed by 2.5 times. Numerical comparisons are available in Tables 10.4 (Inference Time column) and 10.5. In Table 10.5, the difference in training time between the SAM 1 and SAM 2 variants highlights the advantage of using BFloat16 compared to Float32. BFloat16 uses 16

Table 10.4

Quantitative evaluation results on the test set. The top accuracy (*Acc*), precision (*P*), recall (*R*), F1-score, and AP for each version are indicated in bold.

Model	Image Encoder	Parameters	Trainable	*Acc*	*P*	*R*	F1-Score	AP	Inference Time (Img/s)
SAM 1	TinyViT	10.130 M	10.123 M	0.805	0.869	0.916	0.892	0.805	17 (64*)
	ViT-Base	93.735 M	93.729 M	**0.891**	**0.920**	**0.966**	**0.942**	**0.910**	8 (8)
	LoRA(ViT-L)	312.408 M	4.123 M	0.848	0.900	0.936	0.918	0.867	3 (4)
	LoRA(ViT-H)	641.172 M	4.140 M	0.833	0.884	0.936	0.909	0.838	2 (4)
SAM 2	Hiera-T	38.945 M	38.939 M	0.865	0.906	0.951	0.928	0.878	23 (32)
	Hiera-S	46.043 M	46.037 M	0.771	0.847	0.897	0.871	0.783	21 (32)
	Hiera-B+	80.833 M	80.827 M	**0.889**	**0.932**	**0.951**	**0.941**	**0.898**	19 (32)
	LoRA(Hiera-L)	224.432 M	11.723 M	0.739	0.834	0.867	0.850	0.758	15 (32)

* Inference batch size.

Table 10.5

Training time comparison.

	SAM 1				SAM 2			
	TinyViT	ViT-B	LoRA(ViT-L)	LoRA(ViT-H)	Hiera-T	Hiera-S	Hiera-B+	LoRA (Hiera-L)
Total Time	1 h 25 min	6 h 36 min	7 h 21 min	8 h 21 min	1 h	1 h 5 min	1 h 21 min	34 min
Avg. Time Per Fold	17 min	1 h 19 min	1 h 28 min	1 h 40 min	12 min	13 min	16 min	7 min

bits instead of 32 bits (FP32), halving memory usage and enabling more operations per second. Moreover, it highlights the difference in training time for each variant of each model, based on the total parameters that need to be trained and the differences in the architecture of Hiera compared to ViT, which makes it more efficient. These differences are explained in Section 10.2.2.1.

Table 10.6 presents the detailed defect detection performance for each variant, including *TP*s, *FP*s, and *FN*s. To demonstrate the lack of knowledge in SAM 1 and SAM 2, their performance on the test set without any prior knowledge of B-scan images is provided in Table 10.7, where their poor performance strongly necessitates their fine-tuning.

Figure 10.7 presents the training and validation loss curves for various variants, with the epoch corresponding to the best checkpoint clearly marked by a purple star. Figure 10.8 showcases the outputs of the SAM 1 and SAM 2 variants on three representative B-scan images from the test set, providing a visual comparison of their performance.

10.4.2 ABLATION STUDY

In this section, three aspects are explored: fine-tuning only the mask decoder, the effect of reducing training data, and the effect of using cross-entropy loss in combination with Dice loss on the performance of the fine-tuned model.

Table 10.6

True positive (*TP*), false positive (*FP*), and false negative (*FN*) values for each variant of SAM 1 and SAM 2 on the test set. The top-performing variant of each model is indicated in bold.

Model	Image Encoder	*TP*	*FP*	*FN*
SAM 1	TinyViT	186	28	17
	ViT-B	196	17	7
	LoRA(ViT-L)	190	21	13
	LoRA(ViT-H)	190	25	13
SAM 2	Hiera-T	193	20	10
	Hiera-S	182	33	21
	Hiera-B+	193	14	10
	LoRA(Hiera-L)	176	35	27

Table 10.7

Performance of SAM 1 and SAM 2 without any fine-tuning on the ultrasonic B-scan images evaluated on the test set. The top-performing variant of each model is indicated in bold.

Model	Image Encoder	*TP*	*FP*	*FN*	F1-Score	AP
SAM 1	TinyViT	0	11	203	0	0
	ViT-B	12	94	191	0.078	0.009
	LoRA(ViT-L)	4	265	199	0.017	0
	LoRA(ViT-H)	10	353	193	0.035	0.002
SAM 2	Hiera-T	6	41	197	0.048	0.007
	Hiera-S	15	70	188	0.104	0.013
	Hiera-B+	3	10	200	0.028	0.006
	LoRA(Hiera-L)	2	7	201	0.019	0.004

10.4.2.1 Why Not Just Fine-Tune the Mask Decoder?

To ensure that fine-tuning both the image encoder and mask decoder blocks of SAMs is not excessive and that the mask decoder alone cannot adequately handle defect detection in B-scan images, the results of experiments where only the mask decoder of the top-performing variant of each model is fine-tuned are presented in Table 10.8. The configurations for these two experiments are the same as those reported in Table 10.4, except that the image encoder is frozen.

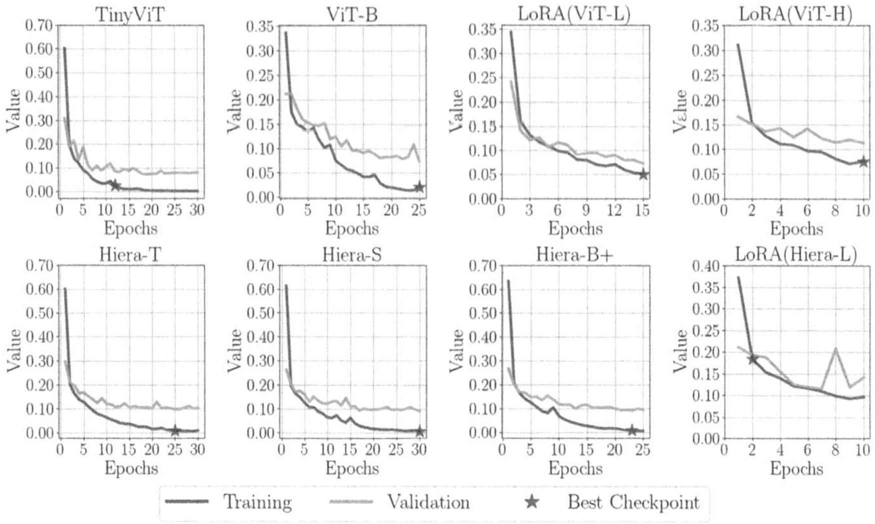

Figure 10.7 Comparison of training and validation losses for SAM variants, with SAM 1 (**top row**) and SAM 2 (**bottom row**) using different image encoders.

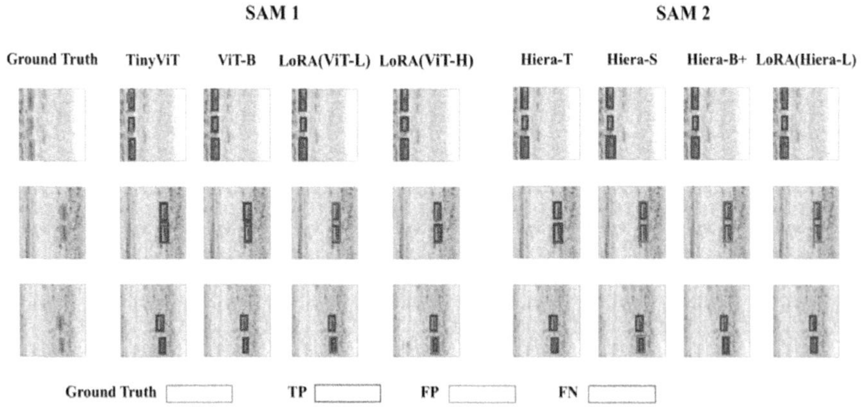

Figure 10.8 Comparison of the performance of each variant of SAM 1 and SAM 2 on three images from the test set.

10.4.2.2 Training Data Size vs. Performance

To highlight the trade-off between training data availability and model effectiveness, top-performing variants are fine-tuned with fewer annotations and evaluated on the test set. The amount of training annotations is scaled by factors ranging from 0.5 to 0.9 (in steps of 0.1) per fold, while the validation and test sets remain unchanged. Figure 10.9 compares SAM 1 with ViT-B and SAM 2 with Hiera-B+ across various

Table 10.8

Performance of the top-performing SAM 1 and SAM 2 variants after repeating the fine-tuning process with only the mask decoder. The ↑ and ↓ arrows indicate the change compared to the results reported in Table 10.4, where both the mask decoder and the image encoder are fine-tuned.

Model	Image Encoder	Trainable	*TP*	*FP*	*FN*	*Acc*	*P*	*R*	F1-Score	AP
SAM 1	ViT-B	4.058 M	159 (−37 ↓)	135 (+118 ↑)	44 (+37 ↑)	0.470 (−47.3% ↓)	0.541 (−41.2% ↓)	0.783 (−18.9% ↓)	0.640 (−32.1% ↓)	0.447 (−50.9% ↓)
SAM 2	Hiera-B+	11.721 M	143 (−50 ↓)	228 (+214 ↑)	60 (+50 ↑)	0.332 (−62.7% ↓)	0.385 (−58.7% ↓)	0.704 (−26.0% ↓)	0.498 (−47.1% ↓)	0.289 (−67.8% ↓)

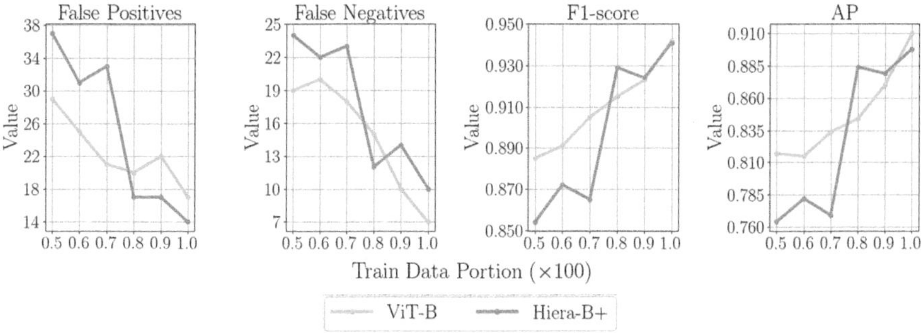

Figure 10.9 Comparing *FP*, *FN*, and performance metrics for top-performing SAM 1 and SAM 2 variants using varying training data portions.

training data scale factors, evaluating *FP*s, *FN*s, F1-score, and the AP metric. With approximately half the training annotations, the F1-score of the top-performing variant of SAM 1 decreases by 6%, while that of SAM 2 decreases by 9.2%. The number of *FN*s increases by 171% (from 7 to 19) for ViT-B and by 140% (from 10 to 24) for Hiera-B+.

10.4.2.3 Cross-Entropy Loss Analysis

Cross-entropy measures the difference between two distributions [84], which is computed by

$$L_{\text{CE}} = -\frac{1}{N}\sum_{i=1}^{N}\sum_{c=1}^{C} g_i^c \log s_i^c \tag{10.8}$$

where g_i^c indicates whether class label c is the correct classification for pixel i, and s_i^c represents the corresponding predicted probability. In this chapter, which focuses on a single class of defects, the equation simplifies to

$$L_{\text{CE}} = -\frac{1}{N}\sum_{i=1}^{N} g_i \log s_i \tag{10.9}$$

Cross-entropy loss is added, with a λ scale, to the Dice loss in Equation 10.7 to

Table 10.9

Performance of top-performing SAM 1 and SAM 2 after considering cross-entropy loss in addition to Dice loss. The ↑ and ↓ arrows indicate the change compared to the results reported in Table 10.6, where only Dice loss is considered during the fine-tuning process.

Model	Image Encoder	λ	*TP*	*FP*	*FN*	*Acc*	*P*	*R*	F1-Score	AP
SAM 1	ViT-B	0.5	198 (+2 ↑)	15 (−2 ↓)	5 (−2 ↓)	0.908 (+1.9% ↑)	0.930 (+1.1% ↑)	0.975 (+0.9% ↑)	0.952 (+1.1% ↑)	0.922 (+1.3% ↑)
SAM 2	Hiera-B+	0.6	196 (+3 ↑)	11 (−3 ↓)	7 (−3 ↓)	0.916 (+3.0% ↑)	0.947 (+1.6% ↑)	0.966 (+1.6% ↑)	0.956 (+1.6% ↑)	0.934 (+4.0% ↑)

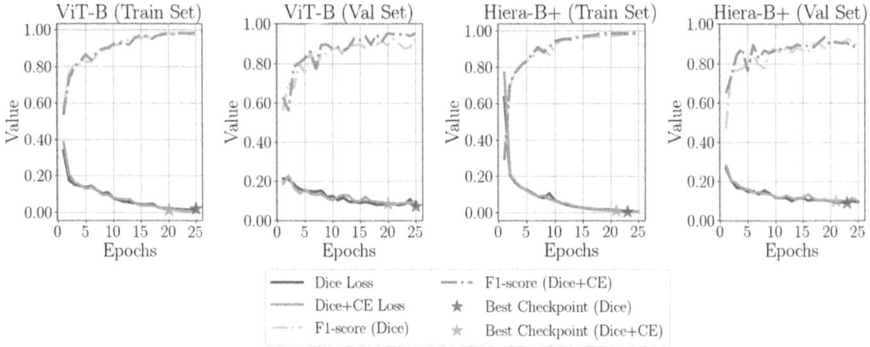

Figure 10.10 Comparison of the effect of adding a cross-entropy loss term on the loss and F1-score trends for both the training set and validation set of the top-performing SAM 1 and SAM 2 variants.

explore its effectiveness on the top-performing variant of each model, resulting in

$$L_{\text{Total}} = L_{\text{Dice}} + \lambda L_{\text{CE}} \tag{10.10}$$

To do so, λ is set from 0.1 to 1.0 with a step of 0.1, and the fine-tuning process is repeated for SAM 1 with ViT-B and SAM 2 with Hiera-B+ using the same configurations described in 10.3.5.3. The λ value at which an improvement is observed, compared to using only Dice loss for fine-tuning, is reported in Table 10.9, along with the performance results. Cross-entropy loss can enhance the model's ability to reduce errors in misclassifying the background as a defect. Figure 10.10 compares the training and validation losses, as well as the F1-score performance on the training and validation sets, with and without the cross-entropy loss term added to the Dice loss. As can be seen, the addition of cross-entropy loss also leads to reaching the best checkpoint faster than the default scenario.

10.4.3 LIMITATIONS AND FUTURE WORKS

Limitations: This chapter presents the application of deploying SAM as an AI-assistant for interpreting real-world ultrasonic B-scan images in AUT. However, some aspects remain to be explored before achieving this goal. For instance, inferring directly from original image sizes, rather than converting them to 64×64 patches, could reduce pre-processing and post-processing steps, although this approach would require more training data to achieve state-of-the-art performance. Additionally, incorporating human feedback could enable the model to learn from false positives and negatives, further improving its performance.

Future Works: Future tasks focus on enhancing both the dataset and model customization to improve performance and efficiency. Regarding the enhancement of the model's generalization, including more defects would improve the proposed method's generalization. However, the human experts inspected only the LOF, which has the greatest impact on weld quality in the studied dataset. By leveraging state-of-the-art models for generating synthetic image data, such as stable diffusion [89], the dataset will be expanded with high-quality images, significantly reducing the time and energy required for data collection. This will facilitate the development of an image encoder pre-trained specifically on ultrasonic B-scan images, utilizing novel self-supervision methods. Moreover, situations where geometric welds are hardly distinguishable from defects, even for expert human operators, should be studied as sources of uncertainty to explore the model's performance. Simultaneously, other architectures, such as mamba vision [90], will be investigated to further refine and customize the model for specific needs.

10.5 CONCLUSIONS

In this chapter, a vision foundation model, SAM, is utilized to design an AI-assistant tool for interpreting AUT data. To achieve this, the image encoder and mask decoder of four different variants of SAM 1 and SAM 2 are fine-tuned on a proprietary ultrasonic B-scan image dataset containing a single class of defect. The experiments are conducted using a stratified five-fold cross-validation approach with both vanilla and LoRA fine-tuning methods. The proposed method leverages SAM in a fully automated and promptable manner for localizing defects in B-scan images, eliminating the need for confidence thresholding and NMS, which are commonly employed in deep learning-based object detection models. Additionally, the effects of dataset size on the model's performance and the inclusion of cross-entropy loss alongside the Dice loss term are studied. The proposed method, by utilizing SAM models, can address some of the difficulties associated with deploying previous deep learning-based methods studied in prior research [35] on the same dataset. These challenges include memory usage, scalability, and inference speed for low-power devices, which are widely used in these industries.

The assessment indicates that using SAM for automated weld defect detection can enhance inspection efficiency and reduce the workload of human inspectors in industries employing AUT for NDT. Furthermore, companies developing NDT tools should consider offering open-source software to facilitate seamless integration of AI models with existing acquisition systems.

REFERENCES

1. LeCun, Y.; Bengio, Y.; Hinton, G. Deep learning. *Nature* **2015**, *521*, 436–444.
2. Ajmi, C.; Zapata, J.; Elferchichi, S.; Laabidi, K. Advanced Faster-RCNN Model for Automated Recognition and Detection of Weld Defects on Limited X-Ray Image Dataset. *J. Nondestruct. Eval.* **2024**, *43*, 14.
3. Totino, B.; Spagnolo, F.; Perri, S. RIAWELC: A Novel dataset of radiographic images for automatic weld defects classification. *Int. J. Electr. Comput. Eng. Res.* **2023**, *3*, 13–17.
4. Naddaf-Sh, S.; Naddaf-Sh, M.M.; Zargarzadeh, H.; Dalton, M.; Ramezani, S.; Elpers, G.; Baburao, V.S.; Kashani, A.R. Real-Time Explainable Multiclass Object Detection for Quality Assessment in 2-Dimensional Radiography Images. *Complexity* **2022**, *2022*, 4637939.
5. Naddaf-Sh, M.M.; Naddaf-Sh, S.; Zargarzadeh, H.; Zahiri, S.M.; Dalton, M.; Elpers, G.; Kashani, A.R. 9 - Defect detection and classification in welding using deep learning and digital radiography. In *Fault Diagnosis and Prognosis Techniques for Complex Engineering Systems*; Karimi, H., Ed.; Academic Press: Cambridge, MA, USA, 2021; pp. 327–352. https://doi.org/10.1016/B978-0-12-822473-1.00007-0.
6. Kim, Y.H.; Lee, J.R. Automated data evaluation in phased-array ultrasonic testing based on A-scan and feature training. *NDT E Int.* **2024**, *141*, 102974. https://doi.org/10.1016/j.ndteint.2023.102974.
7. Pyle, R.J.; Bevan, R.L.; Hughes, R.R.; Rachev, R.K.; Ali, A.A.S.; Wilcox, P.D. Deep learning for ultrasonic crack characterization in NDE. *IEEE Trans. Ultrason. Ferroelectr. Freq. Control* **2020**, *68*, 1854–1865.
8. Medak, D.; Posilović, L.; Subašić, M.; Budimir, M.; Lončarić, S. Automated defect detection from ultrasonic images using deep learning. *IEEE Trans. Ultrason. Ferroelectr. Freq. Control* **2021**, *68*, 3126–3134.
9. Ye, J.; Ito, S.; Toyama, N. Computerized ultrasonic imaging inspection: From shallow to deep learning. *Sensors* **2018**, *18*, 3820.
10. Block, S.B.; Da Silva, R.D.; Lazzaretti, A.E.; Minetto, R. LoHi-WELD: A novel industrial dataset for weld defect detection and classification, a deep learning study, and future perspectives. *IEEE Access* **2024**, *12*, 77442–77453.
11. Qi, H.; Cheng, L.; Kong, X.; Zhang, J.; Gu, J. WDLS: Deep Level Set Learning for Weakly Supervised Aeroengine Defect Segmentation. *IEEE Trans. Ind. Inform.* **2024**, *20*, 303–313. https://doi.org/10.1109/TII.2023.3261889.
12. Tu, X.L.; Zhang, J.; Gambaruto, A.M.; Wilcox, P.D. A framework for computing directivities for ultrasonic sources in generally anisotropic, multi-layered media. *Wave Motion* **2024**, *128*, 103299.
13. Swornowski, P.J. Scanning of the internal structure part with laser ultrasonic in aviation industry. *Scanning* **2011**, *33*, 378–385.
14. Dwivedi, S.K.; Vishwakarma, M.; Soni, A. Advances and researches on non destructive testing: A review. *Mater. Today Proc.* **2018**, *5*, 3690–3698.
15. Cantero-Chinchilla, S.; Croxford, A.J.; Wilcox, P.D. Optimising laser-induced phased-arrays for defect detection in continuous inspections. *NDT E Int.* **2024**, *144*, 103091. https://doi.org/10.1016/j.ndteint.2024.103091.
16. Xie, L.; Lian, Y.; Du, F.; Wang, Y.; Lu, Z. Optical methods of laser ultrasonic testing technology in the industrial and engineering applications: A review. *Opt. Laser Technol.* **2024**, *176*, 110876. https://doi.org/10.1016/j.optlastec.2024.110876.

17. Davis, G.; Stratoudaki, T.; Lukacs, P.; Riding, M.W.; Al Fuwaires, A.; Kamintzis, P.; Pieris, D.; Keenan, A.; Wilcox, P.; Pierce, G.; et al. Near-surface defect detection in additively manufactured components using laser induced phased arrays with surface acoustic wave crosstalk suppression. *Mater. Des.* **2023**, *236*, 112453. https://doi.org/10.1016/j.matdes.2023.112453.
18. Krizhevsky, A.; Sutskever, I.; Hinton, G.E. ImageNet Classification with Deep Convolutional Neural Networks. In *Proceedings of the Advances in Neural Information Processing Systems*; Pereira, F., Burges, C., Bottou, L., Weinberger, K., Eds.; Curran Associates, Inc.: Newry, UK, 2012; Volume 25.
19. Posilović, L.; Medak, D.; Subašić, M.; Petković, T.; Budimir, M.; Lončarić, S. Flaw Detection from Ultrasonic Images using YOLO and SSD. In Proceedings of the 2019 11th International Symposium on Image and Signal Processing and Analysis (ISPA), Dubrovnik, Croatia, 23–25 September 2019; pp. 163–168. https://doi.org/10.1109/ISPA.2019.8868929.
20. Medak, D.; Posilović, L.; Subašić, M.; Budimir, M.; Lončarić, S. DefectDet: A deep learning architecture for detection of defects with extreme aspect ratios in ultrasonic images. *Neurocomputing* **2022**, *473*, 107–115. https://doi.org/10.1016/j.neucom.2021.12.008.
21. Tan, M.; Pang, R.; Le, Q.V. EfficientDet: Scalable and Efficient Object Detection. *arXiv* **2020**, arXiv:1911.09070. https://doi.org/10.48550/arXiv.1911.09070.
22. Virkkunen, I.; Koskinen, T.; Jessen-Juhler, O.; Rinta-Aho, J. Augmented ultrasonic data for machine learning. *J. Nondestruct. Eval.* **2021**, *40*, 4.
23. Goodfellow, I.; Pouget-Abadie, J.; Mirza, M.; Xu, B.; Warde-Farley, D.; Ozair, S.; Courville, A.; Bengio, Y. Generative adversarial nets. *Adv. Neural Inf. Process. Syst.* **2014**, *27*, 2672–2680.
24. Posilović, L.; Medak, D.; Subašić, M.; Budimir, M.; Lončarić, S. Generative adversarial network with object detector discriminator for enhanced defect detection on ultrasonic B-scans. *Neurocomputing* **2021**, *459*, 361–369.
25. Posilović, L.; Medak, D.; Subašić, M.; Budimir, M.; Lončarić, S. Generating ultrasonic images indistinguishable from real images using Generative Adversarial Networks. *Ultrasonics* **2022**, *119*, 106610. https://doi.org/10.1016/j.ultras.2021.106610.
26. Posilović, L.; Medak, D.; Milković, F.; Subašić, M.; Budimir, M.; Lončarić, S. Deep learning-based anomaly detection from ultrasonic images. *Ultrasonics* **2022**, *124*, 106737. https://doi.org/10.1016/j.ultras.2022.106737.
27. Akcay, S.; Atapour-Abarghouei, A.; Breckon, T.P. GANomaly: Semi-Supervised Anomaly Detection via Adversarial Training. *arXiv* **2018**, arXiv:1805.06725. https://doi.org/10.48550/arXiv.1805.06725.
28. Defard, T.; Setkov, A.; Loesch, A.; Audigier, R. PaDiM: A Patch Distribution Modeling Framework for Anomaly Detection and Localization. *arXiv* **2020**, arXiv:2011.08785. https://doi.org/10.48550/arXiv.1805.06725.
29. Rudolph, M.; Wandt, B.; Rosenhahn, B. Same Same However, DifferNet: Semi-Supervised Defect Detection with Normalizing Flows. *arXiv* **2020**, arXiv:2008.12577. https://doi.org/10.48550/arXiv.2008.12577.
30. Milković, F.; Filipović, B.; Subašić, M.; Petković, T.; Lončarić, S.; Budimir, M. Ultrasound Anomaly Detection Based on Variational Autoencoders. In Proceedings of the 2021 12th International Symposium on Image and Signal Processing and Analysis

(ISPA), Zagreb, Croatia, 13–15 September 2021; pp. 225–229. https://doi.org/10.1109/ISPA52656.2021.9552041.

31. Kingma, D.P. Auto-encoding variational bayes. *arXiv* **2013**, arXiv:1312.6114.
32. Shi, X.; Chen, Z.; Wang, H.; Yeung, D.Y.; Wong, W.K.; Woo, W.c. Convolutional LSTM network: A machine learning approach for precipitation nowcasting. *Adv. Neural Inf. Process. Syst.* **2015**, *28*, 802–810.
33. Medak, D.; Posilović, L.; Subašić, M.; Budimir, M.; Lončarić, S. Deep Learning-Based Defect Detection From Sequences of Ultrasonic B-Scans. *IEEE Sensors J.* **2022**, *22*, 2456–2463. https://doi.org/10.1109/JSEN.2021.3134452.
34. Ye, J.; Toyama, N. Automatic defect detection for ultrasonic wave propagation imaging method using spatio-temporal convolution neural networks. *Struct. Health Monit.* **2022**, *21*, 2750–2767. https://doi.org/10.1177/14759217211073503.
35. Naddaf-Sh, A.M.; Baburao, V.S.; Zargarzadeh, H. Automated Weld Defect Detection in Industrial Ultrasonic B-Scan Images Using Deep Learning. *NDT* **2024**, *2*, 108–127. https://doi.org/10.3390/ndt2020007.
36. Zhu, X.; Su, W.; Lu, L.; Li, B.; Wang, X.; Dai, J. Deformable detr: Deformable transformers for end-to-end object detection. *arXiv* **2020**, arXiv:2010.04159.
37. Jocher, G.; Chaurasia, A.; Qiu, J. Ultralytics YOLOv8. *J. Comput. Commun.* **2023**, *11*. (accessed on 27 November 2023) https://github.com/ultralytics/ultralytics.
38. Bommasani, R.; Hudson, D.A.; Adeli, E.; Altman, R.; Arora, S.; von Arx, S.; Bernstein, M.S.; Bohg, J.; Bosselut, A.; Brunskill, E.; et al. On the opportunities and risks of foundation models. *arXiv* **2021**, arXiv:2108.07258.
39. Achiam, J.; Adler, S.; Agarwal, S.; Ahmad, L.; Akkaya, I.; Aleman, F.L.; Almeida, D.; Altenschmidt, J.; Altman, S.; Anadkat, S.; et al. Gpt-4 technical report. *arXiv* **2023**, arXiv:2303.08774.
40. Dubey, A.; Jauhri, A.; Pandey, A.; Kadian, A.; Al-Dahle, A.; Letman, A.; Mathur, A.; Schelten, A.; Yang, A.; Fan, A.; et al. The llama 3 herd of models. *arXiv* **2024**, arXiv:2407.21783.
41. Reid, M.; Savinov, N.; Teplyashin, D.; Lepikhin, D.; Lillicrap, T.; Alayrac, J.b.; Soricut, R.; Lazaridou, A.; Firat, O.; Schrittwieser, J.; et al. Gemini 1.5: Unlocking multimodal understanding across millions of tokens of context. *arXiv* **2024**, arXiv:2403.05530.
42. Ramesh, A.; Pavlov, M.; Goh, G.; Gray, S.; Voss, C.; Radford, A.; Chen, M.; Sutskever, I. Zero-shot text-to-image generation. In Proceedings of the International Conference on Machine Learning. PMLR, Online, 18–24 July 2021; pp. 8821–8831.
43. Wang, X.; Zhang, X.; Cao, Y.; Wang, W.; Shen, C.; Huang, T. Seggpt: Segmenting everything in context. *arXiv* **2023**, arXiv:2304.03284.
44. Kirillov, A.; Mintun, E.; Ravi, N.; Mao, H.; Rolland, C.; Gustafson, L.; Xiao, T.; Whitehead, S.; Berg, A.C.; Lo, W.Y.; et al. Segment anything. In Proceedings of the IEEE/CVF International Conference on Computer Vision, Paris, France, 1–6 October 2023; pp. 4015–4026.
45. Ravi, N.; Gabeur, V.; Hu, Y.T.; Hu, R.; Ryali, C.; Ma, T.; Khedr, H.; Rädle, R.; Rolland, C.; Gustafson, L.; et al. Sam 2: Segment anything in images and videos. *arXiv* **2024**, arXiv:2408.00714.
46. Ma, J.; He, Y.; Li, F.; Han, L.; You, C.; Wang, B. Segment anything in medical images. *Nat. Commun.* **2024**, *15*, 654.

47. Mou, L.; Zhao, Y.; Fu, H.; Liu, Y.; Cheng, J.; Zheng, Y.; Su, P.; Yang, J.; Chen, L.; Frangi, A.F.; et al. Segment Anything Model for Medical Images? *Med. Image Anal.* **2023**, *92*, 103061.
48. Deng, R.; Cui, C.; Liu, Q.; Yao, T.; Remedios, L.W.; Bao, S.; Landman, B.A.; Wheless, L.E.; Coburn, L.A.; Wilson, K.T.; et al. Segment Anything Model (SAM) for Digital Pathology: Assess Zero-shot Segmentation on Whole Slide Imaging. *arXiv* **2023**, arXiv:2304.04155.
49. Zhang, Y.; Shen, Z.; Jiao, R. Segment anything model for medical image segmentation: Current applications and future directions. *Comput. Biol. Med.* **2024**, *171*, 108238. https://doi.org/10.1016/j.compbiomed.2024.108238.
50. Osco, L.P.; Wu, Q.; de Lemos, E.L.; Goncalves, W.N.; Ramos, A.P.M.; Li, J.; Marcato, J. The Segment Anything Model (SAM) for remote sensing applications: From zero to one shot. *Int. J. Appl. Earth Obs. Geoinf.* **2023**, *124*, 103540. https://doi.org/10.1016/j.jag.2023.103540.
51. Cao, Y.; Xu, X.; Sun, C.; Cheng, Y.; Du, Z.; Gao, L.; Shen, W. Segment any anomaly without training via hybrid prompt regularization. *arXiv* **2023**, arXiv:2305.10724.
52. Hu, B.; Gao, B.; Tan, C.; Wu, T.; Li, S.Z. Segment anything in defect detection. *arXiv* **2023**, arXiv:2311.10245.
53. Ahmadi, M.; Lonbar, A.G.; Sharifi, A.; Beris, A.T.; Nouri, M.; Javidi, A.S. Application of segment anything model for civil infrastructure defect assessment. *arXiv* **2023**, arXiv:2304.12600.
54. Ding, H.; Gao, J.; Yuan, Y.; Wang, Q. SamLP: A Customized Segment Anything Model for License Plate Detection. *arXiv* **2024**, arXiv:2401.06374.
55. Dosovitskiy, A. An image is worth 16×16 words: Transformers for image recognition at scale. *arXiv* **2020**, arXiv:2010.11929.
56. Wu, K.; Zhang, J.; Peng, H.; Liu, M.; Xiao, B.; Fu, J.; Yuan, L. Tinyvit: Fast pretraining distillation for small vision transformers. In Proceedings of the European Conference on Computer Vision, Tel Aviv, Israel, 23–27 October 2022; Springer: Cham, Switzerland, 2022; pp. 68–85.
57. He, K.; Chen, X.; Xie, S.; Li, Y.; Dollár, P.; Girshick, R. Masked autoencoders are scalable vision learners. In Proceedings of the IEEE/CVF Conference on Computer Vision and Pattern Recognition, New Orleans, LA, USA, 18–24 June 2022; pp. 16000–16009.
58. Li, Y.; Mao, H.; Girshick, R.; He, K. Exploring plain vision transformer backbones for object detection. In Proceedings of the European Conference on Computer Vision, Tel Aviv, Israel, 23–27 October 2022; Springer: Cham, Switzerland, 2022; pp. 280–296.
59. Vaswani, A. Attention is all you need. *Adv. Neural Inf. Process. Syst.* **2017**,*30*, 5998–6008.
60. Ba, J.L. Layer normalization. *arXiv* **2016**, arXiv:1607.06450.
61. Zhang, C.; Han, D.; Qiao, Y.; Kim, J.U.; Bae, S.H.; Lee, S.; Hong, C.S. Faster Segment Anything: Towards Lightweight SAM for Mobile Applications. *arXiv* **2023**, arXiv:2306.14289.
62. Howard, A.; Sandler, M.; Chu, G.; Chen, L.C.; Chen, B.; Tan, M.; Wang, W.; Zhu, Y.; Pang, R.; Vasudevan, V.; et al. Searching for mobilenetv3. In Proceedings of the IEEE/CVF International Conference on Computer Vision, Seoul, Republic of Korea, 27 October–2 November 2019; pp. 1314–1324.
63. Hendrycks, D.; Gimpel, K. Gaussian error linear units (gelus). *arXiv* **2016**, arXiv:1606.08415.

64. Ioffe, S. Batch normalization: Accelerating deep network training by reducing internal covariate shift. *arXiv* **2015**, arXiv:1502.03167.
65. Ryali, C.; Hu, Y.T.; Bolya, D.; Wei, C.; Fan, H.; Huang, P.Y.; Aggarwal, V.; Chowdhury, A.; Poursaeed, O.; Hoffman, J.; et al. Hiera: A hierarchical vision transformer without the bells-and whistles. In Proceedings of the International Conference on Machine Learning, PMLR, Honolulu, HI, USA, 23–29 July 2023; pp. 29441–29454.
66. Li, Y.; Wu, C.Y.; Fan, H.; Mangalam, K.; Xiong, B.; Malik, J.; Feichtenhofer, C. Mvitv2: Improved multiscale vision transformers for classification and detection. In Proceedings of the IEEE/CVF Conference on COMPUTER vision and Pattern Recognition, New Orleans, LA, USA, 18–24 June 2022; pp. 4804–4814.
67. Bolya, D.; Ryali, C.; Hoffman, J.; Feichtenhofer, C. Window Attention is Bugged: How not to Interpolate Position Embeddings. *arXiv* **2023**, arXiv:2311.05613.
68. Lin, T.Y.; Dollár, P.; Girshick, R.; He, K.; Hariharan, B.; Belongie, S. Feature pyramid networks for object detection. In Proceedings of the IEEE Conference on Computer Vision and Pattern Recognition, Honolulu, HI, USA, 21–26 July 2017; pp. 2117–2125.
69. Radford, A.; Kim, J.W.; Hallacy, C.; Ramesh, A.; Goh, G.; Agarwal, S.; Sastry, G.; Askell, A.; Mishkin, P.; Clark, J.; et al. Learning transferable visual models from natural language supervision. In Proceedings of the International Conference on Machine Learning, PMLR, Online, 18–24 July 2021; pp. 8748–8763.
70. Azizi, S.; Culp, L.; Freyberg, J.; Mustafa, B.; Baur, S.; Kornblith, S.; Chen, T.; MacWilliams, P.; Mahdavi, S.S.; Wulczyn, E.; et al. Robust and efficient medical imaging with self-supervision. *arXiv* **2022**, arXiv:2205.09723.
71. Ding, N.; Qin, Y.; Yang, G.; Wei, F.; Yang, Z.; Su, Y.; Hu, S.; Chen, Y.; Chan, C.M.; Chen, W.; et al. Parameter-efficient fine-tuning of large-scale pre-trained language models. *Nat. Mach. Intell.* **2023**, *5*, 220–235.
72. Dutt, R.; Ericsson, L.; Sanchez, P.; Tsaftaris, S.A.; Hospedales, T. Parameter-efficient fine-tuning for medical image analysis: The missed opportunity. *arXiv* **2023**, arXiv:2305.08252.
73. Aghajanyan, A.; Zettlemoyer, L.; Gupta, S. Intrinsic dimensionality explains the effectiveness of language model fine-tuning. *arXiv* **2020**, arXiv:2012.13255.
74. Hu, E.J.; Shen, Y.; Wallis, P.; Allen-Zhu, Z.; Li, Y.; Wang, S.; Wang, L.; Chen, W. Lora: Low-rank adaptation of large language models. *arXiv* **2021**, arXiv:2106.09685.
75. Gu, H.; Dong, H.; Yang, J.; Mazurowski, M.A. How to build the best medical image segmentation algorithm using foundation models: A comprehensive empirical study with Segment Anything Model. *arXiv* **2024**, arXiv:2404.09957.
76. Zhang, K.; Liu, D. Customized Segment Anything Model for Medical Image Segmentation. *arXiv* **2023**, arXiv:2304.13785.
77. Wada, K. Labelme: Image Polygonal Annotation with Python. Available online: `https://zenodo.org/records/5711226` (accessed on 28 June 2024).
78. Paszke, A.; Gross, S.; Massa, F.; Lerer, A.; Bradbury, J.; Chanan, G.; Killeen, T.; Lin, Z.; Gimelshein, N.; Antiga, L.; et al. Pytorch: An imperative style, high-performance deep learning library. *Adv. Neural Inf. Process. Syst.* **2019**, *32*, 8026–8037.
79. Russakovsky, O.; Deng, J.; Su, H.; Krause, J.; Satheesh, S.; Ma, S.; Huang, Z.; Karpathy, A.; Khosla, A.; Bernstein, M.; et al. ImageNet Large Scale Visual Recognition Challenge. *Int. J. Comput. Vis. (IJCV)* **2015**, *115*, 211–252. `https://doi.org/10.1007/s11263-015-0816-y`.

80. van der Walt, S.J.; Schönberger, J.L.; Nunez-Iglesias, J.; Boulogne, F.; Warner, J.D.; Yager, N.; Gouillart, E.; Yu, T.; the scikit-image contributors. scikit-image: Image processing in Python. *PeerJ* **2014**, *2*, e453. https://doi.org/10.7717/peerj.453.
81. Zhang, Y.; Ni, Q. A Novel Weld-Seam Defect Detection Algorithm Based on the S-YOLO Model. *Axioms* **2023**, *12*, 697. https://doi.org/10.3390/axioms12070697.
82. Milletari, F.; Navab, N.; Ahmadi, S.A. V-net: Fully convolutional neural networks for volumetric medical image segmentation. In Proceedings of the 2016 Fourth International Conference on 3D Vision (3DV), Stanford, CA, USA, 25–28 October 2016; IEEE: Piscataway, NJ, USA, 2016; pp. 565–571.
83. Isensee, F.; Jaeger, P.F.; Kohl, S.A.; Petersen, J.; Maier-Hein, K.H. nnU-Net: A self-configuring method for deep learning-based biomedical image segmentation. *Nat. Methods* **2021**, *18*, 203–211.
84. Ma, J.; Chen, J.; Ng, M.; Huang, R.; Li, Y.; Li, C.; Yang, X.; Martel, A.L. Loss Odyssey in Medical Image Segmentation. *Med. Image Anal.* **2021**, *71*, 102035. https://doi.org/10.1016/j.media.2021.102035.
85. Cardoso, M.J.; Li, W.; Brown, R.; Ma, N.; Kerfoot, E.; Wang, Y.; Murrey, B.; Myronenko, A.; Zhao, C.; Yang, D.; et al. Monai: An open-source framework for deep learning in healthcare. *arXiv* **2022**, arXiv:2211.02701.
86. Zhu, Y.; Shen, Z.; Zhao, Z.; Wang, S.; Wang, X.; Zhao, X.; Shen, D.; Wang, Q. MeLo: Low-rank Adaptation is Better than Fine-tuning for Medical Image Diagnosis. *arXiv* 2023, arXiv:2311.08236. https://doi.org/10.48550/arXiv.2311.08236.
87. Mangrulkar, S.; Gugger, S.; Debut, L.; Belkada, Y.; Paul, S.; Bossan, B. PEFT: State-of-the-Art Parameter-Efficient Fine-Tuning Methods. 2022 Available online: https://github.com/huggingface/peft (accessed on 1 October 2024).
88. Ma, J.; Kim, S.; Li, F.; Baharoon, M.; Asakereh, R.; Lyu, H.; Wang, B. Segment anything in medical images and videos: Benchmark and deployment. *arXiv* **2024**, arXiv:2408.03322.
89. Rombach, R.; Blattmann, A.; Lorenz, D.; Esser, P.; Ommer, B. High-resolution image synthesis with latent diffusion models. In Proceedings of the IEEE/CVF Conference on Computer Vision and Pattern Recognition, New Orleans, LA, USA, 18–24 June 2022; pp. 10684–10695.
90. Zhu, L.; Liao, B.; Zhang, Q.; Wang, X.; Liu, W.; Wang, X. Vision mamba: Efficient visual representation learning with bidirectional state space model. *arXiv* **2024**, arXiv:2401.09417.

11 Synthetic Data-Driven Weld Defect Classification

Amir-M. Naddaf-Sh, Vinay S Baburao, Zina Ben-Miled, Hassan Zargarzadeh

CHAPTER SUMMARY

Automated ultrasonic testing (AUT) serves as a vital method for evaluating critical infrastructure in industries such as oil and gas. However, a significant challenge in deploying artificial intelligence (AI)-based interpretation methods for AUT data lies in improving their reliability and effectiveness, particularly due to the inherent scarcity of real-world defective data. This study directly addresses data scarcity in a weld defect classification task, specifically concerning the detection of lack of fusion (LOF) defects in weld inspections using a proprietary industrial ultrasonic B-scan image dataset. This chapter leverages state-of-the-art generative models, including Generative Adversarial Networks (GANs) and Denoising Diffusion Probabilistic Models (DDPM) (StyleGAN3, VQGAN with an unconditional transformer, and Stable Diffusion), to produce realistic B-scan images depicting LOF defects. The fine-tuned Transformer-based models, including ViT-Base, Swin-Tiny, and MobileViT-Small classifiers, on the regular B-scan image dataset are then applied to retain only high-confidence positive synthetic samples from each method. The impact of these synthetic images on the classification performance of a ResNet-50 model is evaluated, where it is fine-tuned with cumulative additions of synthetic images, ranging from 10 to 200 images. Its accuracy on the test set increases by 38.9% relative to the baseline with the addition of either 80 synthetic images using VQGAN with an unconditional transformer or 200 synthetic images by StyleGAN3 to the training set, and by 36.8% with the addition of 150 synthetic images by Stable Diffusion. This also outperforms Transformer-based vision models that are trained on regular training data. Concurrently, knowledge distillation experiments involve training ResNet-50 as a student model, leveraging the expertise of ViT-Base and Swin-Tiny as teacher models to demonstrate the effectiveness of adding the synthetic data to the training set, where the greatest enhancement is observed to be 34.7% relative to the baseline. This work contributes to advancing robust, AI-assisted tools for critical infrastructure inspection and offers practical pathways for enhancing available models in resource-constrained industrial environments.

11.1 INTRODUCTION

Non-destructive testing (NDT) methods, such as X-radiography, Ultrasonic testing (UT), Magnetic particle testing, and Eddy current testing, are critical for maintaining

DOI: 10.1201/9781003766575-11

the safety and integrity of components across safety-critical industries (e.g., aerospace, power generation, oil and gas) by enabling the early detection of defects and preventing economic losses and environmental disasters [1]. The labor-intensive nature of NDT and the high risk of human error necessitate the development of automated assistive inspection tools. Deep learning-based solutions for automated defect detection in digital X-radiography [2, 4, 3] and UT data [5, 6] are crucial for mitigating the risks and costs associated with maintaining aging infrastructure. Specifically, UT is favored for its ability to detect small subsurface flaws and generate internal structural images [8, 7, 9], and is often utilized in high-precision Automated Ultrasonic Testing (AUT) systems, where data is presented in formats like A-scans, B-scans (preferred for detailed visual representation), and C-scans. While Artificial Intelligence, particularly deep learning, offers powerful tools for transforming AUT through automatic feature extraction and pattern recognition, its real-world application is significantly challenged by the limited availability of defective data from industrial settings [10].

To address data scarcity (not limited to AUT) in automated defect detection generally, the authors in [10] present a two-phase approach for generating synthetic, self-labeled defective industrial images. Using generative models like Stable Diffusion, this method aims to overcome data scarcity in automated defect detection. First, the model learns the concept of industrial images. Second, it learns to control the image generation process to produce images with specific defect characteristics. This synthetic data significantly improves crack segmentation model performance, particularly when real annotated data is limited. In another study, ref. [11] proposed the "In&Out" approach, which leverages diffusion models for realistic in-distribution data augmentation in surface defect detection. This method combines diffusion-generated samples with traditional out-of-distribution samples, significantly improving classification performance, especially when positive samples are scarce. DIAG (Diffusion-based In-distribution Anomaly Generation) [12], a training-free pipeline leveraging latent diffusion models for data augmentation in surface defect detection, implements a human-in-the-loop approach in which domain experts provide multimodal guidance through text descriptions and region localization of anomalies, enhancing interpretability and enabling plausible anomaly generation without real positive data.

In AUT, specifically for extending the B-scan image dataset, expensive data augmentation using virtual flaws [13] enables the training of modern deep convolutional neural networks for flaw detection in phased-array ultrasonic data, achieving human-level performance in defect classification. DetectionGAN [14], a novel deep learning Generative Adversarial Network (GAN) [15], experimentally demonstrates that expanding training datasets by generating synthetic B-scan images with defects at distinct locations significantly improves the performance of deep convolutional neural object detection networks for defect detection. A key contribution is the use of a pre-trained object detector as an additional discriminator within the GAN, which enhances the quality of generated images and ensures accurate defect placement. DiffUT [16] addresses the challenge of limited sample availability in ultrasonic

B-scan wheel defect data for high-speed rail by using a novel diffusion model, which learns probability and noise distributions through the diffusion process and generates synthetic data that significantly improves defect detection performance.

In this study, the scarcity of data for weld defect classification as a binary task, along with the poor performance of a deep convolutional neural network model, ResNet-50, on the defect class in industrial B-scan images, is addressed by leveraging synthetic data generation to enhance classification performance. In previous work [17], the current authors introduced a real-world ultrasonic B-scan image dataset collected by UT experts during automated girth weld inspection in oil and gas pipelines, focusing on the detection of lack of fusion (LOF) defects. They also evaluated the baseline performance of state-of-the-art deep learning models. In their subsequent study, [18] leveraged a foundational vision model for weld defect detection in industrial ultrasonic B-scan images, demonstrating improved performance with an F1-Score of nearly 0.940 on the same dataset and reducing implementation complexity compared to previously studied methods.

The main contributions of this chapter are as follows:

- State-of-the-art generative models, including StyleGAN3, VQGAN with an unconditional transformer, and Stable Diffusion, are fine-tuned to synthesize realistic B-scan images of LOF defects under limited data conditions.
- High-confidence synthetic defect images are selected through a filtering process using fine-tuned Transformer-based classifiers.
- The impact of synthetic augmentation on ResNet-50 is demonstrated, showing substantial performance improvement where the baseline model fails.
- A comparison between synthetic augmentation and knowledge distillation is provided, demonstrating that synthetic data yields greater performance gains for this task.

The remainder of this chapter is organized as follows. The Section 11.2 explains the theoretical background and key terms, along with details about the image generation methods, the deep learning models used, and the evaluation metrics. The Experimental Setup Section 11.3 describes the dataset, the implementation of the experiments, and the hyperparameter configurations. The Section 11.4 presents a performance comparison of ResNet-50 models fine-tuned on the B-scan dataset, incorporating synthetic images generated by different image generation methods. Finally, the Section 11.5 summarizes the chapter.

11.2 MATERIALS AND METHODS

11.2.1 LACK OF FUSION (LOF) DETECTION VIA ULTRASONIC B-SCAN

A B-scan in ultrasonic testing provides a two-dimensional image by compiling A-scan waveforms collected along an inspection path, mapping acoustic data to a coordinate system where one axis is the probe's position and the other is time-of-flight (depth) [19]. Pixel intensities represent the amplitude of reflected echoes, which are used to identify material boundaries and internal flaws. Specifically, Lack of Fusion

(LOF) is a common, critical weld defect appearing in a B-scan as a distinct reflector echo, often elongated, at the non-fused interface. Interpreting these B-scan indications requires expert knowledge to distinguish subtle defect signals from background noise and geometry echoes, a task that is challenging to automate. Figure 11.1A summarizes the general weld inspection and B-scan acquisition process. This study focuses on detecting LOF in B-scans as a binary classification problem (defect vs. non-defect), addressing the difficulty of classifying these low-contrast or noise-masked indications. Figure 11.2 illustrates sample B-scan images from the dataset containing LOF defects.

11.2.2 SYNTHETIC IMAGE GENERATION

Synthetic Image Generation refers to the production of artificial yet realistic images via generative modeling. In this study, additional B-scan images that simulate LOF defects are created using advanced generative models. By learning the statistical features of real B-scan defect images, new examples that mimic real defect appearances can be created by these models. The motivation is to augment the limited real dataset with diverse synthetic examples of defects, addressing class imbalance (far fewer defect images than non-defect images) and improving the deep learning-based image classifiers performance. In this chapter, three generative approaches are utilized:

11.2.2.1 StyleGAN3

A state-of-the-art Generative Adversarial Network (GAN) developed by NVIDIA (NVIDIA Corporation, Santa Clara, CA, USA) [20], renowned for its ability to generate high-quality, natural-looking images. A key innovation in StyleGAN3 is its alias-free architecture, which addresses aliasing artifacts common in previous GANs by ensuring every layer of the synthesis network produces a continuous signal, leading to more natural transformations. This is achieved by carefully redesigning the up-sampling and down-sampling layers with proper anti-aliasing filters. Furthermore, StyleGAN3 interprets all signals within the network as continuous, rather than discrete pixel values, which allows for true sub-pixel equivariance to translation and rotation. This fundamental change prevents "texture sticking," where details appear fixed to pixel coordinates instead of moving naturally with the depicted object. The internal representations of StyleGAN3 develop their own coordinate systems, ensuring that fine details are correctly attached to underlying surfaces.

11.2.2.2 Denoising Diffusion Probabilistic Models (DDPMs)

DDPMs [21] are a class of generative models that learn to reverse a gradual noising process. They progressively transform random noise into structured data, such as images, over a sequence of steps, achieving high-quality sample generation. Stable Diffusion is a prominent latent diffusion model that builds upon these DDPM principles, operating in a compressed latent space for efficiency. In this study, to fine-tune Stable Diffusion [22] on a limited amount of data, two techniques are utilized:

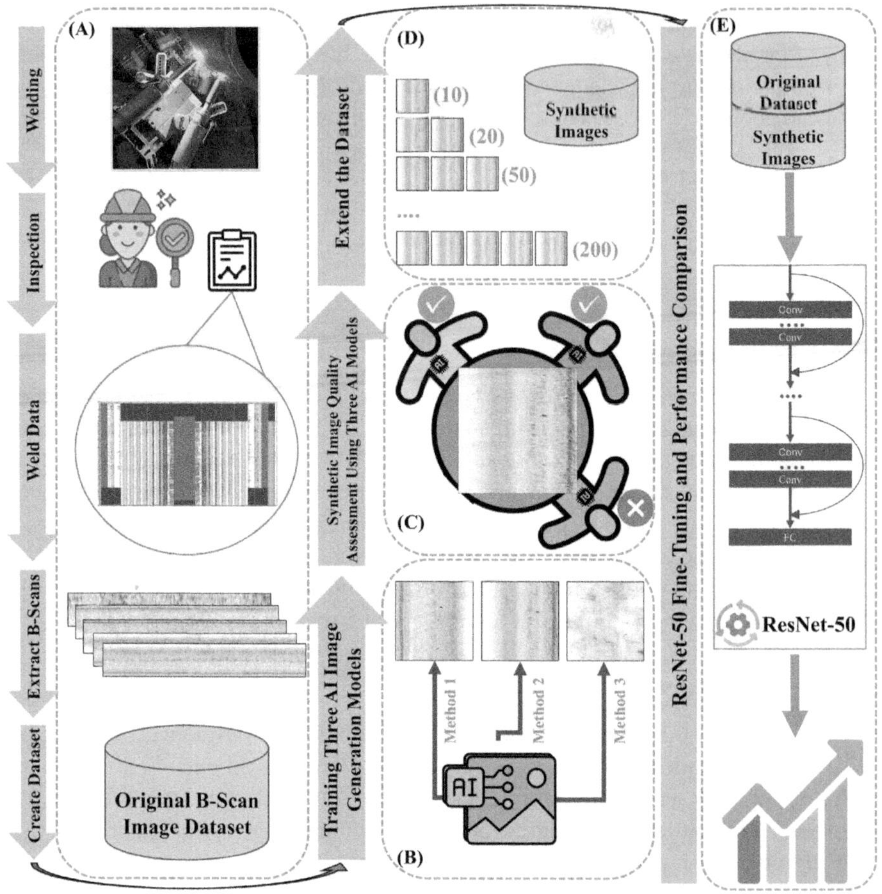

Figure 11.1 (**A**) Acquisition of weld inspection data and extraction of B-scan strip charts to form the original dataset. (**B**) Training of three synthetic image generation models using the positive LOF samples. (**C**) Quality assessment of synthetic images using three Transformer-based classifiers to retain only high-confidence samples. (**D**) Progressive extension of the training dataset by adding filtered synthetic LOF images in controlled increments. (**E**) Fine-tuning of the ResNet-50 using the extended dataset and comparison of performance across all experimental conditions.

DreamBooth [23] and LoRA [24].DreamBooth is a fine-tuning technique that enables text-to-image diffusion models to generate personalized images of a specific subject using only a few reference photos. It maintains the subject's consistent appearance by employing a unique identifier token alongside a class noun, such as "a photo of [V] cat." This approach allows for diverse image generation while effectively preserving the model's pre-trained knowledge. Meanwhile, LoRA (Low-Rank Adaptation) is an efficient fine-tuning technique that adds small, trainable low-rank

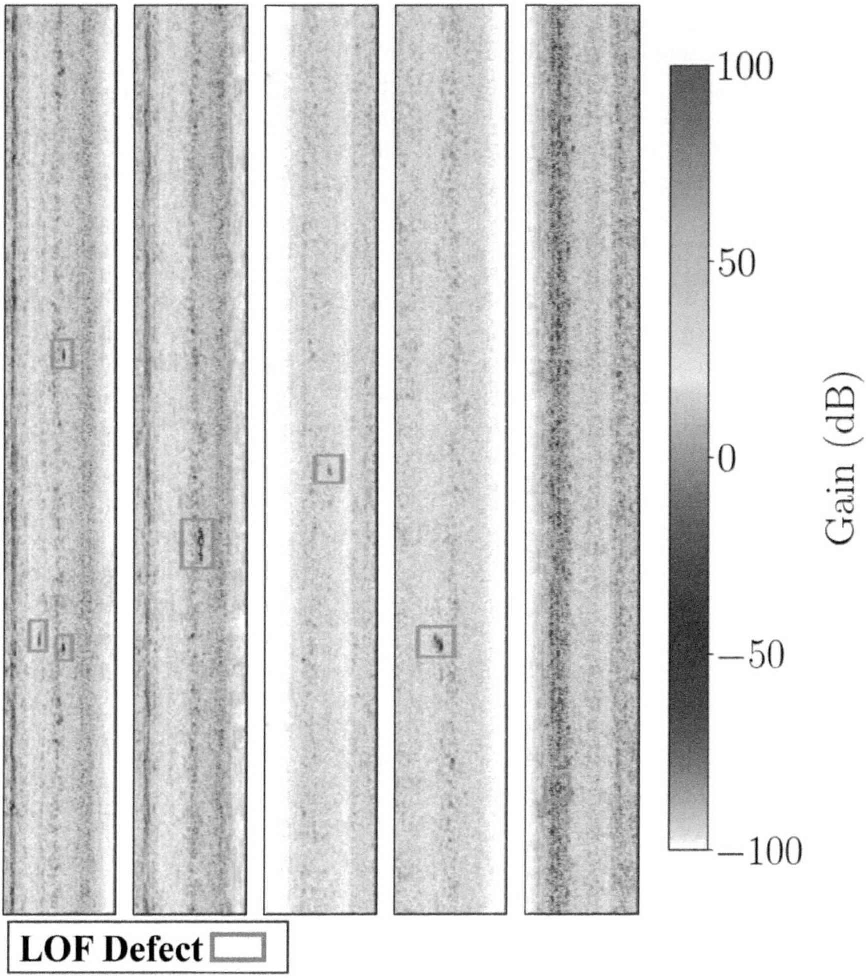

Figure 11.2 Samples of B-scan images with LOF defect from the dataset.

matrices to the model's layers, which drastically reduces memory and computational demands compared to full fine-tuning.

11.2.2.3 VQGAN with an Unconditional Transformer

Vector Quantized Generative Adversarial Network (VQGAN) [25] combined with an unconditional transformer is a two-stage generative model. It operates by encoding images into a compact latent space using a vector quantized autoencoder, which discretizes the latent representations into a finite codebook of embeddings. This latent space significantly reduces the dimensionality of the data while preserving semantic structure, allowing for efficient manipulation and high-quality reconstruction of images. The GAN component ensures that generated images from the latent codes are photorealistic and detailed, overcoming the blurriness often seen in traditional autoencoders. Subsequently, an autoregressive transformer models the composition of these visual tokens, enabling high-resolution image synthesis. The synthetic image generation process illustrated in Figure 11.1B.

11.2.3 DEEP LEARNING ARCHITECTURES FOR IMAGE CLASSIFICATION

Deep learning, a subset of machine learning, employs multi-layered artificial neural networks to learn hierarchical representations from data. In the context of image classification, these models automatically extract and learn relevant features from images, enabling them to identify and categorize visual data with high accuracy. The deep learning models utilized in this study are briefly introduced below:

11.2.3.1 ResNet-50

ResNet-50 [26] is a prominent Convolutional Neural Network (CNN) architecture, known for its deep structure facilitated by residual connections or skip connections. These connections enable the training of very deep networks by mitigating the vanishing gradient problem, allowing for effective feature learning. In this study, ResNet-50 was selected as the primary classification model to observe the influence of synthetic image inclusion in the training set. The choice of ResNet-50 as the target model for performance improvement is strategic; it is a well-established and computationally less demanding CNN compared to more recent Transformer architectures. Improving its performance via synthetic data makes it a more practical and deployable choice for real-world industrial settings, such as edge devices or mobile inspection units, where computational resources are often limited.

11.2.3.2 Vision Transformers

In recent years, transformer architectures [27], originally developed for sequence modeling in natural language processing, have been adapted for image recognition tasks. Vision Transformers (ViT) split an image into patches and employ self-attention mechanisms to learn global contextual relationships, in contrast to CNNs that learn local spatial filters. ViT-Base [28] directly applies a standard transformer encoder to sequences of image patches. It is pre-trained on vast amounts of image data, enabling strong generalization capabilities. Swin-Tiny [29] is a hierarchical Vision Transformer that computes self-attention within non-overlapping windows, offering improved efficiency and scalability compared to global attention. It addresses the computational cost of ViTs while maintaining strong performance. MobileViT [30] integrates efficient Transformer layers into a lightweight CNN for mobile-friendly deployment. In this study, all three Transformer-based models are utilized for filtering the generated synthetic images, while only ViT-Base and Swin-Tiny are used for knowledge distillation experiments. Figure 11.1C illustrates the synthetic image quality assessment step, where three Transformer-based classifiers filter generated samples based on high-confidence LOF predictions. Figure 11.1D depicts the incremental dataset extension process in which the filtered synthetic LOF images are added to the training set in controlled quantities.

11.2.4 KNOWLEDGE DISTILLATION

Knowledge distillation (KD) [31] is a machine learning technique designed to transfer learned representations and knowledge from a large, often complex and high-performing teacher model to a smaller, more computationally efficient student model. Unlike traditional training, where a model learns from hard labels, KD often involves the student learning from soft targets (pseudo-probabilities) or intermediate representations generated by the teacher model, thereby mimicking the teacher's decision-making process and often achieving comparable performance. In this chapter, KD is employed to provide a comparative benchmark against the synthetic data augmentation strategy, demonstrating the effectiveness of the proposed method in enhancing the performance of a weaker model (ResNet-50).

11.2.5 EVALUATION METRICS

To quantitatively evaluate defect classification performance, standard classification metrics are employed. Accuracy is calculated as the sum of true positives (*TP*s) and true negatives (*TN*s) divided by the total number of test samples, providing the proportion of correctly classified defect and non-defect instances. However, accuracy can be misleading in the presence of class imbalance for example, if defects are rare, a model that always predicts "no defect" may still achieve high accuracy. Therefore, greater emphasis is placed on precision, recall, and the F1-Score, particularly for the defect class.

Precision (*P*) is computed as the number of *TP*s divided by the sum of *TP*s and false positives (*FP*s). This metric reflects the proportion of predicted defects that are actual defects, thus indicating the model's ability to control false alarms (*FP*s). Recall (*R*) is determined by dividing the number of *TP*s by the sum of *TP*s and false negatives (*FN*s), capturing the model's ability to identify actual defects. The F1-Score, representing the harmonic mean of precision and recall, provides a balanced measure of performance. A perfect F1-Score of 1.0 implies complete detection of defects with no *FP*s or *FN*s.

These metrics are reported on a held-out test set of real B-scan images. In practical applications, high precision is valued to avoid unnecessary follow-up actions on *FP*s, while high recall is crucial to minimize missed defect cases. The F1-Score serves as a concise indicator of the model's overall effectiveness in detecting the rare defect class. AUC (Area Under the Receiver Operating Characteristic Curve) is a comprehensive metric that evaluates the model's ability to distinguish between positive and negative classes across all possible classification thresholds. It is robust to class imbalance and provides an aggregate measure of performance. Figure 11.1E shows the final stage of the workflow, where the augmented dataset is used to fine-tune ResNet-50 and evaluate its performance.

$$\text{Accuracy} = \frac{TP+TN}{TP+FP+FN+TN}$$

$$\text{Precision} = \frac{TP}{TP+FP}$$

$$\text{Recall} = \frac{TP}{TP+FN}$$

$$\text{F1-Score} = \frac{2 \times P \times R}{P+R}$$

11.3 EXPERIMENTAL SETUP

11.3.1 DATASET

The dataset utilized in this study is a proprietary collection of ultrasonic B-scan images, compiled from 87 distinct weld inspection records. These records were acquired by experienced UT experts using an automated girth weld inspection system, which employs phased array technology for the inspection of onshore oil and gas pipelines. It is important to note that the authors of this study were not involved in the initial inspection process; rather, the B-scan images were collected from pre-existing proprietary weld inspection data provided by these UT experts. This utilization of real-world B-scan images with genuine defects contrasts with the prevalent reliance in research on in-laboratory images or artificially flawed data, highlighting the data acquisition challenges faced by industry in training robust deep learning-based models.

The recorded weld data represents J-bevel weld types, and the primary weld defect class identified and targeted for detection is lack of fusion (LOF), recognized as the most common defect encountered in automated girth welding within this domain. Each recorded weld originally consisted of a 1920 × 1080 image containing multiple strip charts, which represent the output from the channels of the phased array probes used in AUT. These strip charts also provided views from both downstream and upstream phased array probes, along with detailed indications of the weld zones represented by each channel.

From these weld inspection records, a total of 359 B-scan strip chart images were collected by the authors. The dataset exhibits a class imbalance, comprising 234 negative images (no LOF defect) and 95 positive images (containing the lack of fusion defect). The distribution of the B-scan dataset includes images predominantly from the Fill 1, Hot Pass 1, and Hot Pass 2 zones of the weld. Figure 11.3 illustrates these weld zones. It is meant to show a general idea of how the weld is built up in different regions, Root, Body, and Cap, using multiple passes. The number of passes shown (like Root, LCP, Hot-Pass 1, etc.) is just an example. In actual practice, the number of weld passes will vary depending on factors like the pipe's wall thickness.

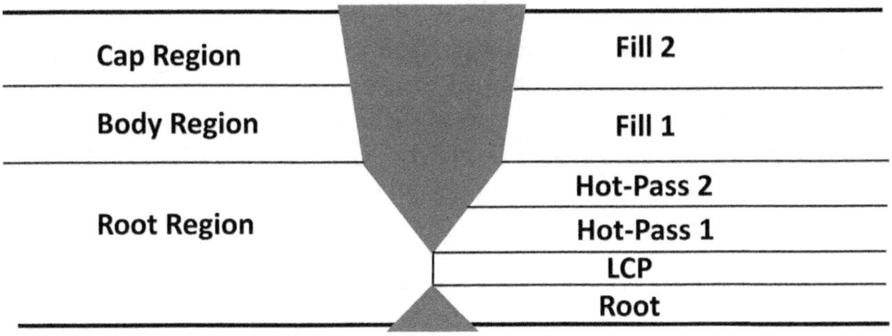

Figure 11.3 Weld Zones.

Thicker pipes typically require more fill and cap passes The core task addressed in this study is image classification, specifically identifying the presence or absence of LOF defects. For robust experimental evaluation, the dataset was split into training, validation, and test sets using a stratified sampling approach. This stratification ensured that the proportion of positive and negative samples was maintained across the splits, with a ratio of 0.64 for training, 0.16 for validation, and 0.20 for testing. For transparency, it is acknowledged that not using a weld-ID–based grouped split may introduce a potential limitation; however, because each weld is represented by strip charts originating from different weld zones, the risk of information leakage is considered to be substantially reduced. For more information about the dataset, please refer to [17]. Table 11.1 shows the number of negative and positive B-scan images in each split.

11.3.2 EXPERIMENTS IMPLEMENTATION DETAILS

11.3.2.1 Fine-Tuning Procedures for Synthetic Image Generation Models

To address the scarcity of positive (LOF defect) samples in the proprietary dataset, three advanced generative models, VQGAN with an unconditional transformer, Stable Diffusion, and StyleGAN3, are employed to generate synthetic B-scan images. These models are fine-tuned exclusively on the positive images from the training set

Table 11.1
Number of negative (no LOF defect) and positive (with LOF defect) B-scan images in each subset of the dataset (train, validation, and test sets).

Type of Image	Train	Validation	Test
Negative	155	39	49
Positive	73	19	24
Total	228	58	73

of the original dataset. Specifically, 73 positive images, which are resized to 256 × 256 pixels, are oversampled through duplication to a total of 146 images to provide a sufficient base for fine-tuning the generative models. A random seed of 42 is consistently used across all experiments to ensure reproducibility. All experiments are implemented in PyTorch 2.0.2 (StyleGAN3), 1.7.1 (VQGAN), 2.1.2 (Stable Diffusion) [32] and executed on a single NVIDIA A100 (40 GB) GPU (NVIDIA Corporation, Santa Clara, CA, USA).

- VQGAN With An Unconditional Transformer: For the VQGAN with an unconditional transformer model, the instructions and default configurations from the main repository (`https://github.com/CompVis/taming-transformers`, accessed on 10 May 2025) are followed. This involves a two-stage fine-tuning process. First, the VQGAN model is fine-tuned on the positive B-scan images for 30 epochs using pre-trained ImageNet weights, with a batch size of 8. The model with the lowest validation loss, which occurred at epoch 24, is selected. Subsequently, an unconditional transformer model, architecturally similar to GPT-2 [33] and configured with the following parameters: vocab_size of 1024, block_size of 512, 24 layers, 16 heads, and an embedding size of 1024, is trained for 20 epochs with a batch size of 32. The best-performing model (based on validation loss), selected from epoch 16, is used for subsequent analysis.
- StyleGAN3: The StyleGAN3 model is fine-tuned using the stylegan3-transform configuration and the checkpoint (stylegan3-t-ffhqu-256x256.pkl) pre-trained on the FFHQ-U 256×256 dataset, adhering to the default settings provided in the official repository (`https://github.com/NVlabs/stylegan3`, accessed on 17 May 2025). Training is carried out for 100,000 iterations with a batch size of 32.
- Stable Diffusion: The fine-tuning procedure for Stable Diffusion follows the guidelines outlined in the open-source repository (https://github.com/kohya-ss/sd-scripts, accessed on 3 May 2025). The sd-v1-5-pruned-noema-fp16 checkpoint (huggingface.co/.../sd-v1-5-pruned-noema-fp16.safetensors, accessed on 3 May 2025) is used as the base model for fine-tuning on the positive B-scan images. DreamBooth fine-tuning, a technique designed to teach a new concept to a diffusion model with fewer iterations and without overwriting its extensive prior knowledge, is applied. For this purpose, 200 regularization images generated using the prompt "background defected" are placed alongside the positive images from the dataset's training set, which are associated with the prompt "skt background defected" (where "skt" serves as a token identifier without semantic meaning). A parameter-efficient fine-tuning (PEFT) approach, specifically LoRA (LoRA-c3Lier), is employed following the configuration of an extended LoRA variant that adapts both linear layers and convolutional layers. In this setup, LoRA adapters are applied to the linear and 1 × 1 convolution projection layers with a rank (`network_dim`) of 32 and an alpha value of 16. In addition, 3 × 3 convolution layers are also adapted using convolutional LoRA

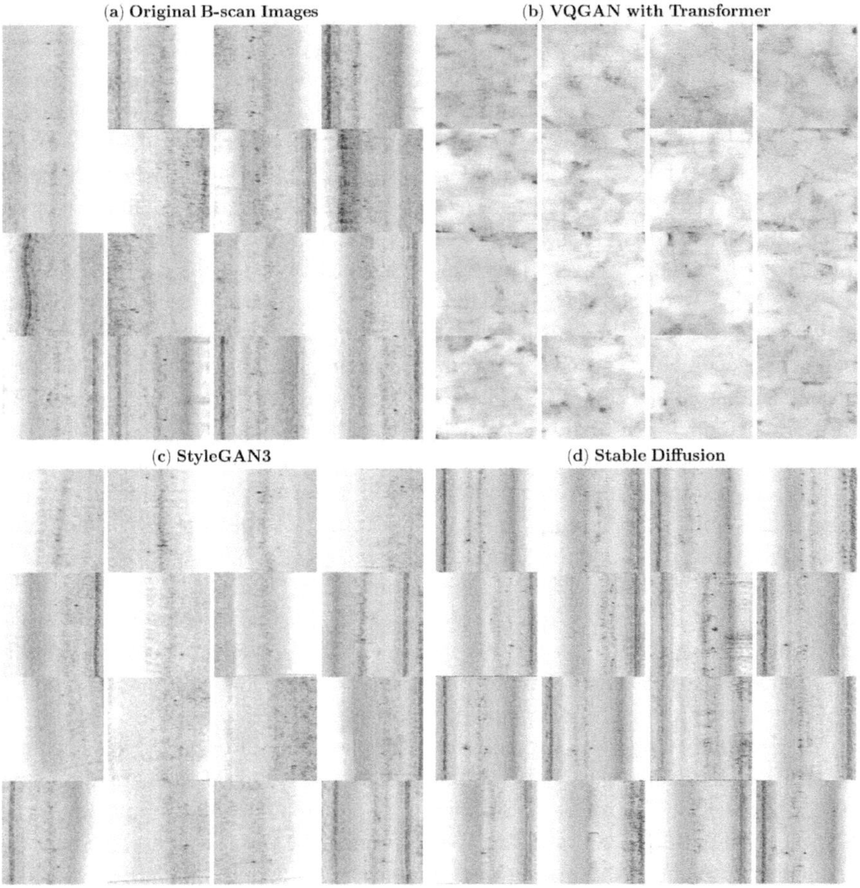

Figure 11.4 Sixteen samples displayed in a 4 × 4 grid for each method: (**a**) B-scan images with LOF defects from the training set; (**b**) defective B-scan images generated using VQGAN with an unconditional transformer; (**c**) defective B-scan images generated using StyleGAN3; and (**d**) defective B-scan images generated using Stable Diffusion.

components configured with rank (`conv_dim`) 4 and alpha 1. Training is conducted for 25 epochs with a learning rate set to 1e-3 and a batch size of 256. Figure 11.4 presents positive sample B-scan images from the training set alongside images generated by each different method.

11.3.2.2 Fine-Tuning Procedures for Image Classification Models

Four different image classification models are fine-tuned on the B-scan image dataset, representing both CNN and Transformer architectures: ResNet-50 (https://huggingface.co/timm/resnet50.a1_in1k, accessed on 30 May 2025), ViT-Base (https://huggingface.co/google/vit-base-patch16-224, accessed on 30 May 2025), Swin Transformer (Tiny) (https://huggingface.co/microsoft/swin-tiny-patch4-window7-224, accessed on 30 May 2025), and MobileViT (Small) (https://huggingface.co/apple/mobilevit-small, accessed on 30 May 2025). All models are initialized with ImageNet-pretrained weights to provide a strong starting point, given the limited data available. The three Transformer-based image classification models are fine-tuned on the original dataset without any synthetic images. These models are chosen for their strong baseline performance on the original dataset and are used as a filtering mechanism for synthetic data. From each fine-tuned synthetic image generation model, 500 synthetic images are generated. From the 500 synthetic images generated by each method, only those that are confidently classified as positive samples by all three high-performing transformer models are retained for inclusion in the training set. This filtering step is considered crucial to ensure that only high-quality, truly representative synthetic positive images are added to the training set. In this way, the inclusion of noisy or ambiguous samples that could negatively impact the classifier's performance is prevented. The specific counts of positively classified synthetic images retained from each method are as follows: 430 images from Stable Diffusion, 211 images from StyleGAN3, and 207 images from VQGAN with an unconditional transformer. The differences in retained samples across the generative methods reflect the quality of the generated images, since all three filters were trained identically and a sample is kept only when all three models agree it is positive.

To evaluate the effect of synthetic image augmentation, ResNet-50 is selected as the primary classification model. The filtered, positively classified synthetic images are randomly added to the original training set in cumulative steps of 10, 20, 50, 80, 100, 120, 150, and 200 images. This incremental approach is used to enable a detailed assessment of the impact of adding synthetic data on model performance. The optimal amount of synthetic data needed and the point at which additional data no longer improves, or may even harm, performance are identified. At each cumulative step, ResNet-50 is fine-tuned on the augmented training set. Figure 11.5 depicts the complete workflow, encompassing the generation of synthetic images, their selection, and the subsequent fine-tuning of Image generation models and ResNet-50.

To provide a comparative analysis and demonstrate the effectiveness of the proposed approach, the knowledge distillation technique is employed. This involves using ViT-Base and Swin-Tiny as teacher models due to their superior performance on the original dataset, while ResNet-50 was designated as the student model. To perform knowledge distillation, the logits (raw prediction scores) from both the teacher model (T) and the student model (S) are first obtained. These logits are then scaled by a temperature parameter, which controls the softness of the teacher's predictions. A higher temperature yields softer probability distributions, encouraging the student to learn the general shape of the teacher's output rather than focusing on exact probabilities. The overall influence of this knowledge distillation process on the student's training is then adjusted by a lambda parameter, which weighs the importance of the

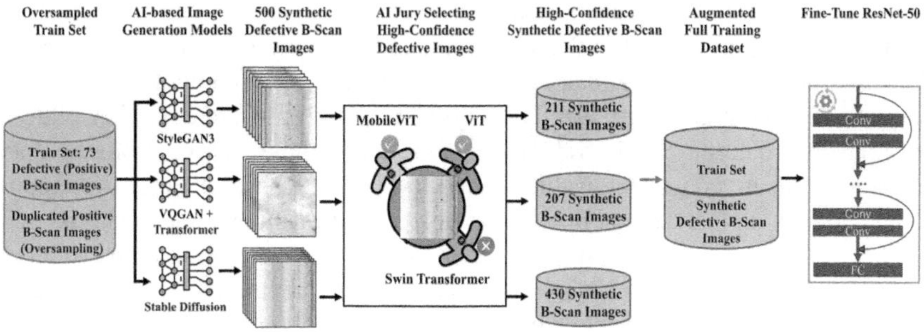

Figure 11.5 The pipeline used for augmenting a limited B-scan image dataset involves AI-based generative models and a selection process to create a high-confidence synthetic image set for fine-tuning ResNet-50. The orange cylinders show original defective B-Scan images in the train set. The blue cylinders show synthetic defective B-Scan images. The purple cylinder shows both defective and non-defective B-Scan images in the train set.

distillation loss. In the experiments, the temperature is set to 5 and lambda to 0.5.

The core of this distillation relies on the Kullback-Leibler (*KL*) Divergence loss [34] to quantify the difference between the probability distribution of the student (*S*) and the teacher (*T*). Given two data distributions, *T* and *S*, the *KL* Divergence measures the additional information required to represent distribution T using distribution *S*. If *T* and *S* are identical, their *KL* divergence is zero, as no extra information is needed to describe *T* from *S*. This characteristic makes *KL* divergence particularly useful in knowledge distillation, as it helps the student model (*S*) to minimize its divergence from the teacher model (*T*), thereby learning the teacher's knowledge. For discrete probability distributions *T* and *S*, the *KL* Divergence of *S* from *T*, denoted as $D_{KL}(T \parallel S)$, is given by:

$$D_{KL}(T \parallel S) = \sum_{x \in X} T(x) \ln\left(\frac{T(x)}{S(x)}\right)$$

where X is set of all possible events and x is an event.

All image classification models involved in the study, including the three Transformer-based models and ResNet-50, are fine-tuned for up to 100 epochs with a batch size of 128. A learning rate of 1×10^{-4} is used, coupled with a cosine learning rate scheduler without weight decay, and 10 warm-up epochs are incorporated to stabilize the initial training. For reproducibility, a fixed random seed of 42 is used across all experiments. Regarding image transformations for each model, in the training set, each image is first randomly cropped and resized to the size specified by the model's input requirements, then randomly flipped horizontally, which only alters visual orientation and is physically valid, and finally normalized using the mean and standard deviation values of the ImageNet dataset. For the validation and test sets, the transformations are deterministic to ensure consistent evaluation: each image is resized, then center-cropped to the model's input size, and finally normalized in the same manner as in the training set. All computational tasks are performed using PyTorch on a single Nvidia A100 (40 GB) GPU. The checkpoint of the model achieving the lowest evaluation loss on the validation set during the fine-tuning process is selected and subsequently evaluated on the test set to compare its classification metrics. The implementation of these experiments is inspired by (HuggingFace Transformers - Image Classification Example, accessed on 30 May 2025) and the following link (Knowledge Distillation Documentation, accessed on 10 June 2025).

11.4 RESULTS AND DISCUSSION

The results of our experiments are summarized in Figures 11.6 and 11.7 and Tables 11.2, 11.3 , 11.4 and 11.5. These illustrate the impact of synthetic data augmentation (using each synthetic image generation method), the effect of incremental synthetic additions, and the outcomes of knowledge distillation. All percentage improvements reported in this work refer to relative increases over the baseline in Table 11.2 unless otherwise specified. Moreover, these tables include AUC and overall accuracy for the binary classification task, along with precision, recall, and F1-Score reported specifically for the defect class.

11.4.1 QUALITATIVE AND FEATURE-SPACE ANALYSIS OF SYNTHETIC IMAGES

To qualitatively evaluate the similarity between real B-scan images and the filtered synthetic defective B-scan images generated by each model, a comparative t-SNE analysis was performed. Feature embeddings were extracted from ViT-Base, Swin-Tiny, and MobileViT-Small for all real positive and negative samples as well as for the filtered synthetic images generated by VQGAN with an unconditional transformer, StyleGAN3, and Stable Diffusion. The resulting feature sets were standardized and projected into two dimensions using t-SNE, enabling a visual comparison of how the synthetic samples align with the distribution of real B-scan images in the learned feature spaces of the classifiers.

The t-SNE projections in Figure 11.6A,B show that VQGAN with an unconditional transformer images formed a distinct, compact cluster within the broader region occupied by the real positive class, like their more abstract visual appearance in Figure 11.4. In contrast, synthetic samples from StyleGAN3 and Stable Diffusion clustered closely with the real defective B-scan images, reflecting their stronger visual resemblance. Importantly, all three synthetic sets remained largely separated

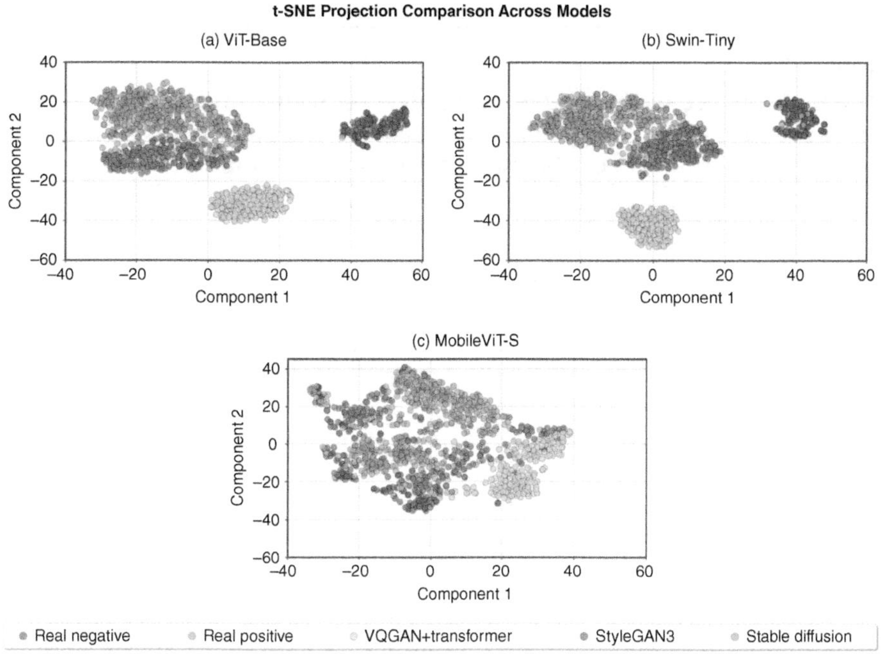

Figure 11.6 t-SNE projections comparing real and synthetic B-scan images in the feature spaces of three Transformer models. Real positive and negative samples form mostly distinct clusters in feature spaces of (**A**) ViT-Base and (**B**) Swin-Tiny. StyleGAN3 and Stable Diffusion synthetic samples overlap substantially with the real positive class, while VQGAN with an unconditional transformer forms a separate cluster, demonstrating a distinct feature representation despite effective defect encoding.

from the real negative class, indicating that the generative models did not introduce substantial misleading artifacts.

These results indicate that, even though visually different, VQGAN with an unconditional transformer images still encode salient low-level characteristics of defective B-scan images, making them useful for model training. The more overlapped appearance of MobileViT-Small embeddings in Figure 11.6C arises from its lightweight architecture, which uses compressed intermediate representations that capture less global structure compared to ViT-Base and Swin-Tiny. This behavior is expected for mobile-oriented models and does not contradict the strong classification performance reported in Table 11.2, because t-SNE often produces more overlapping clusters for compact models and should be interpreted only as a qualitative visualization rather than as direct evidence of poor class separability. In this study, no formal domain-expert evaluation was performed, as the objective was to assess how the synthetic image sets influence the defect classification performance of ResNet-50; accordingly, their representation in the feature spaces of the three Transformer-based

Table 11.2
The performance of the models on the test set without the addition of any synthetic images is presented below. The models were fine-tuned with the same hyperparameters on the training set for a fair comparison.

Model	Params	AUC	Accuracy	Precision	Recall
ResNet-50	25.6 M	0.500	0.671	0.000	0.000
ViT-Base	86.6 M	0.886	0.904	0.870	0.833
Swin-Tiny	28.3 M	0.866	0.877	0.800	0.833
MobileViT-S	5.6 M	0.969	0.918	0.875	0.875

models serves as the qualitative basis for comparison rather than human visual assessment.

11.4.2 RESNET-50 BASELINE PERFORMANCE VS. TRANSFORMER-BASED MODELS

Table 11.2 presents the baseline classification performance of each model (ResNet-50, ViT-Base, Swin-Tiny, MobileViT-Small) trained only on the original dataset. Without augmentation, the ResNet-50 model struggled, especially in recall, precision, and F1-Score, reflecting that it missed all detections from the defect class when trained solely on the limited and imbalanced original dataset; it completely fails to correctly classify any positive samples. It effectively performs no better than random guessing or simply predicting the majority (negative) class. This dramatic baseline failure of ResNet-50 is not just a data point; it is a powerful justification for the entire research endeavor. It demonstrates the critical need for synthetic data to make a standard, deployable CNN like ResNet-50 effective for detecting rare, critical defects in real-world industrial settings.

In contrast, the Transformer-based models demonstrate significantly higher baseline performance on the original dataset. MobileViT-Small, despite being the most lightweight among them (5.6M parameters), achieves a remarkable F1-Score of 0.875 and an AUC of 0.969, highlighting its efficiency and strong generalization capabilities. ViT-Base and Swin-Tiny also exhibit robust performance, establishing them as strong baselines and suitable candidates for teacher models in knowledge distillation.

11.4.3 IMPROVING RESNET-50 FOR LOF DEFECT CLASSIFICATION THROUGH SYNTHETIC DATA

To analyze the relationship between the quantity of synthetic data and model performance, the number of synthetic images in the training set was incrementally increased for fine-tuning ResNet-50. As shown in Figure 11.7, the effect of adding

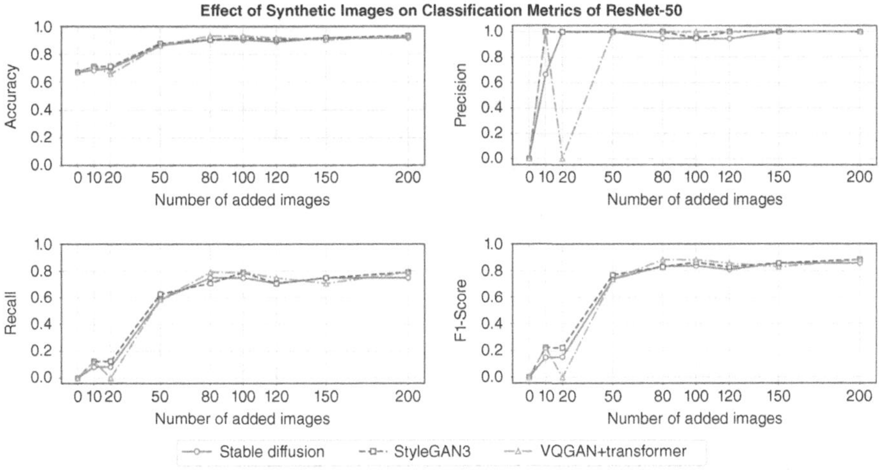

Figure 11.7 Comparison of Accuracy, Precision, Recall, and F1-Score metrics on the test set after adding synthetic images (generated via three different methods) to the training set for fine-tuning ResNet-50.

synthetic images (generated via VQGAN with unconditional transformer, StyleGAN3, and Stable Diffusion) on classification metrics is illustrated by the plots.

Focusing on the F1-Score plot (bottom right), a steep rise in performance is observed with an initial small addition of synthetic images. For all three methods, the F1-Score is seen to climb sharply when moving from 0 to approximately 50 synthetic images. This indicates that even a modest injection of high-quality synthetic data significantly improves the model's learning capability. As more synthetic images are added, the F1-Score continues to improve, but with a tapering slope, eventually plateauing around 100–150 synthetic images. After this point, the improvements become smaller, showing that adding more examples doesn't help the model much anymore. Moreover, no significant degradation in accuracy is observed when synthetic data are added, up to 200 images. This suggests that the filtering and quality control measures for synthetic images are effective in preventing the classifier from being misled.

A robust and consistent trend is observed across all metrics with images generated by StyleGAN3 and Stable Diffusion. Instability is observed with VQGAN with an unconditional transformer when fewer than 50 images are used, as oscillations in the trend are illustrated. However, when synthetic images generated by this model are utilized, ResNet-50 achieves higher performance than the other methods with only 80 images. In Table 11.3, the best performance achieved by each method and the corresponding number of added images are reported. It is important to note that synthetic images generated by the VQGAN with an unconditional transformer method are not like those generated by the other methods or to the original images. However, these images enhance the performance of a deep learning model because they contain

Table 11.3

The best performance achieved by ResNet-50 after adding a specific number (#) of synthetic images generated with three different methods to the training set. The fine-tuning is done with the same hyperparameters as those applied to a regular training set without any synthetic images, ensuring a fair comparison.

Method	# Images	AUC	Accuracy	Precision	Recall	F1-Score
VQGAN + Transformer	80	0.896	0.932	1.000	0.792	0.884
StyleGAN3	200	0.896	0.932	1.000	0.792	0.884
Stable Diffusion	150	0.875	0.918	1.000	0.750	0.857

low-level features of LOF defects, which are represented in the generated images and are helpful for the model, but not for training human experts. If synthetic images are intended to be used for training human experts, only those generated by StyleGAN3 and Stable Diffusion are considered helpful, as their appearance closely resembles that of the original images. Excluding VQGAN with an unconditional transformer, the addition of StyleGAN3-generated images improves the performance of ResNet-50 beyond that of transformer-based models, when 200 synthetic positive samples are used.

The similar best-case performance across the three generative methods in Table 11.3 occurs because each method effectively enriches the positive class with informative synthetic samples. StyleGAN3 and Stable Diffusion generate visually realistic LOF patterns that closely resemble real defects, while VQGAN with an unconditional transformer produces less realistic but still meaningful representations that occupy a complementary region of the feature space. After filtering by the three Transformer-based classifiers, only high-confidence synthetic images are retained, ensuring that all generators provide useful training signals for ResNet-50. Consequently, the model reaches comparable peak performance with each augmentation strategy, even though the number of synthetic samples required to achieve this level differs among the methods.

The diversity analysis in Table 11.4 provides further insight into why StyleGAN3 requires more synthetic images to reach its optimal performance. StyleGAN3 exhibits consistently higher mean pairwise distances (MPD) than VQGAN with an unconditional transformer across all three feature extractors, indicating greater intra-class variability in the generated samples. Its convex hull area (CHA) is also substantially larger than that of VQGAN for Swin-Tiny and MobileViT-Small, reflecting broader coverage of the feature space. This higher variability means that ResNet-50 benefits from a larger number of StyleGAN3 samples before fully internalizing the diversity of the synthesized defect appearances. In contrast, VQGAN-generated samples form tighter, more homogeneous clusters, allowing the model to achieve peak performance with fewer examples, despite their reduced visual realism.

Table 11.4

Diversity metrics for synthetic images generated by the three methods, computed using mean pairwise distance (MPD) and convex hull area (CHA) in the 2-D t-SNE feature space. Values are reported for 200 synthetic samples per generator using embeddings from ViT-Base, Swin-Tiny, and MobileViT-S.

Model	ViT-Base		Swin-Tiny		MobileViT-S	
	MPD	CHA	MPD	CHA	MPD	CHA
VQGAN + Transformer	9.95	569.53	8.42	266.25	15.49	905.52
Stable Diffusion	12.32	862.94	10.59	495.74	17.89	1307.95
StyleGAN3	13.60	488.90	11.07	536.71	24.82	2074.90

Table 11.5

Performance of ResNet-50 as a student model using knowledge distillation. The fine-tuning is done with the same hyperparameters as for a regular training set without synthetic images, ensuring a fair comparison.

Teacher	AUC	Accuracy	Precision	Recall	F1-Score
ViT-Base	0.856	0.836	0.688	0.917	0.786
Swin-Tiny	0.854	0.904	1.000	0.708	0.829

11.4.4 SYNTHETIC DATA VS. KNOWLEDGE DISTILLATION

In Table 11.5, the performance of ResNet-50 after fine-tuning as a student model is provided. Although the enhancement is significant, it cannot achieve the performances achieved by adding synthetic images generated with image generation methods to the training set. With knowledge distillation, the highest performance in terms of F1-Score is achieved when Swin-Tiny is considered as a teacher model, which is 6.63% lower than the performance achieved with the addition of 80 or 200 synthetic images generated via VQGAN with an unconditional transformer or a StyleGAN3, respectively. The reason MobileViT-Small is not utilized for knowledge distillation experiments is that it has significantly fewer parameters as a teacher compared to using ResNet-50 as the student model. Additionally, it produces compressed representations due to its optimization for inference speed and memory efficiency, whereas ResNet-50 can better absorb complex representations from ViT-Base and Swin-Tiny, which have high-dimensional feature representations. Nevertheless, there was no improvement at all in ResNet-50's performance when MobileViT-Small was used as the teacher, compared to its baseline performance. To avoid misinterpretation, it is clarified that the comparison between synthetic data augmentation and knowledge distillation in this study applies only to the ResNet-50 architecture and is not intended to be generalized across other model families.

11.5 CONCLUSIONS

This study addressed the critical issue of data scarcity in automated ultrasonic testing (AUT) for weld defect classification, focusing on the detection of rare and severe lack of fusion (LOF) defects. By leveraging synthetic B-scan image generation, the study demonstrated a substantial improvement in deep learning model performance, particularly for the ResNet-50 architecture. Initially, ResNet-50 exhibits weak performance on the imbalanced real-world dataset for detecting defective B-scan images; its accuracy of 0.671 reflects majority-class accuracy due to the dominance of normal images, resulting in an F1-Score, precision, and recall of 0.000 and an AUC of 0.500. However, when supplemented with synthetic images, specifically 80 generated by VQGAN with an unconditional transformer or 200 generated by StyleGAN3, the model achieved an F1-Score of 0.884 and perfect precision of 1.000, representing up to a 38.9% increase in accuracy on the test set. The study also evaluated different generative models, finding that while VQGAN with an unconditional transformer exhibited some instability with a smaller number of synthetic images, it still enabled high performance with limited synthetic input. StyleGAN3 and Stable Diffusion consistently improved results, with StyleGAN3 achieving similar peak performance using 200 synthetic samples. These findings underscore that high-quality synthetic data, even when visually different from real images, can contribute critical low-level features for effective training. In comparison, knowledge distillation improved baseline performance but remained 6.63% lower relative to the best results achieved using synthetic data. This highlights the superior utility of synthetic data augmentation for training efficient models like ResNet-50, making them practical for deployment in resource-constrained industrial settings such as mobile inspection units or edge devices. Overall, this work advances AI-assisted infrastructure inspection by offering scalable and high-performance solutions using synthetic data.

Although industrial ultrasonic data are used in this study, a complete assessment of operational efficiency would require deployment within an actual AUT inspection workflow. The synthetic image generation steps are carried out entirely offline, and only the final ResNet-50 classifier is employed during inference. As this lightweight model is capable of real-time operation on standard industrial hardware, the proposed approach is considered suitable for integration into existing inspection systems used by UT experts. It is expected that future field deployment will further confirm its practicality and effectiveness in industrial environments. This chapter is part of the first author's Doctor of Engineering dissertation [35] which is being carried out in the Department of Electrical and Computer Engineering at Lamar University.

REFERENCES

1. Adegboye, M.A.; Fung, W.-K.; Karnik, A. Recent Advances in Pipeline Monitoring and Oil Leakage Detection Technologies: Principles and Approaches. *Sensors* **2019**, *19*, 2548. https://doi.org/10.3390/s19112548.

2. Ajmi, C.; Zapata, J.; Elferchichi, S.; Laabidi, K. Advanced Faster-RCNN Model for Automated Recognition and Detection of Weld Defects on Limited X-Ray Image Dataset. *J. Nondestruct. Eval.* **2024**, *43*, 14.
3. Naddaf-Sh, M.M.; Naddaf-Sh, S.; Zargarzadeh, H.; Zahiri, S.M.; Dalton, M.; Elpers, G.; Kashani, A.R. 9—Defect Detection and Classification in Welding Using Deep Learning and Digital Radiography. In *Fault Diagnosis and Prognosis Techniques for Complex Engineering Systems*; Karimi, H., Ed.; Academic Press: Cambridge, MA, USA, 2021; pp. 327–352.
4. Naddaf-Sh, S.; Naddaf-Sh, M.M.; Zargarzadeh, H.; Dalton, M.; Ramezani, S.; Elpers, G.; Baburao, V.S.; Kashani, A.R. Real-Time Explainable Multiclass Object Detection for Quality Assessment in 2-Dimensional Radiography Images. *Complexity* **2022**, *2022*, 4637939.
5. Medak, D.; Posilović, L.; Subašić, M.; Budimir, M.; Lončarić, S. Automated Defect Detection from Ultrasonic Images Using Deep Learning. *IEEE Trans. Ultrason. Ferroelectr. Freq. Control* **2021**, *68*, 3126–3134.
6. Ye, J.; Ito, S.; Toyama, N. Computerized Ultrasonic Imaging Inspection: From Shallow to Deep Learning. *Sensors* **2018**, *18*, 3820.
7. Dwivedi, S.K.; Vishwakarma, M.; Soni, A. Advances and Researches on Non Destructive Testing: A Review. *Mater. Today Proc.* **2018**, *5*, 3690–3698.
8. Swornowski, P.J. Scanning of the Internal Structure Part with Laser Ultrasonic in Aviation Industry. *Scanning* **2011**, *33*, 378–385.
9. Tu, X.L.; Zhang, J.; Gambaruto, A.M.; Wilcox, P.D. A Framework for Computing Directivities for Ultrasonic Sources in Generally Anisotropic, Multi-Layered Media. *Wave Motion* **2024**, *128*, 103299.
10. Valvano, G.; Agostino, A.; De Magistris, G.; Graziano, A.; Veneri, G. Controllable Image Synthesis of Industrial Data Using Stable Diffusion. In Proceedings of the IEEE/CVF Winter Conference on Applications of Computer Vision, Waikoloa, HI, USA, 3–8 January 2024; pp. 5354–5363.
11. Capogrosso, L.; Girella, F.; Taioli, F.; Dalla Chiara, M.; Aqeel, M.; Fummi, F.; Setti, F.; Cristani, M. Diffusion-Based Image Generation for In-Distribution Data Augmentation in Surface Defect Detection. *arXiv* **2024**, arXiv:2406.00501.
12. Girella, F.; Liu, Z.; Fummi, F.; Setti, F.; Cristani, M.; Capogrosso, L. Leveraging Latent Diffusion Models for Training-Free In-Distribution Data Augmentation for Surface Defect Detection. In Proceedings of the 2024 International Conference on Content-Based Multimedia Indexing (CBMI), Reims, France, 4–6 September 2024; IEEE: Piscataway, NJ, USA, 2024; pp. 1–7.
13. Virkkunen, I.; Koskinen, T.; Jessen-Juhler, O.; Rinta-Aho, J. Augmented Ultrasonic Data for Machine Learning. *J. Nondestruct. Eval.* **2021**, *40*, 4.
14. Posilović, L.; Medak, D.; Subašić, M.; Budimir, M.; Lončarić, S. Generative Adversarial Network with Object Detector Discriminator for Enhanced Defect Detection on Ultrasonic B-Scans. *Neurocomputing* **2021**, *459*, 361–369.
15. Goodfellow, I.J.; Pouget-Abadie, J.; Mirza, M.; Xu, B.; Warde-Farley, D.; Ozair, S.; Courville, A.; Bengio, Y. Generative Adversarial Nets. *Adv. Neural Inf. Process. Syst.* **2014**, *27*, 2672–2680.
16. Zhang, Q.; Tian, K.; Zhang, F.; Li, J.; Yang, K.; Luo, L.; Gao, X.; Peng, J. DiffUT: Diffusion-Based Augmentation for Limited Ultrasonic Testing Defects in High-Speed Rail. *NDT E Int.* **2025**, *154*, 103388. `https://doi.org/10.1016/j.ndteint.2025.103388`.

17. Naddaf-Sh, A.M.; Baburao, V.S.; Zargarzadeh, H. Automated Weld Defect Detection in Industrial Ultrasonic B-Scan Images Using Deep Learning. *NDT* **2024**, *2*, 108–127.
18. Naddaf-Sh, A.M.; Baburao, V.S.; Zargarzadeh, H. Leveraging Segment Anything Model (SAM) for Weld Defect Detection in Industrial Ultrasonic B-Scan Images. *Sensors* **2025**, *25*, 277.
19. Krautkrämer, J.; Krautkrämer, H. *Ultrasonic Testing of Materials*; Springer Science and Business Media: Berlin/Heidelberg, Germany, 2013.
20. Karras, T.; Aittala, M.; Laine, S.; Härkönen, E.; Hellsten, J.; Lehtinen, J.; Aila, T. Alias-Free Generative Adversarial Networks. *Adv. Neural Inf. Process. Syst.* **2021**, *34*, 852–863.
21. Ho, J.; Jain, A.; Abbeel, P. Denoising Diffusion Probabilistic Models. *Adv. Neural Inf. Process. Syst.* **2020**, *33*, 6840–6851.
22. Rombach, R.; Blattmann, A.; Lorenz, D.; Esser, P.; Ommer, B. High-Resolution Image Synthesis with Latent Diffusion Models. In Proceedings of the IEEE/CVF Conference on Computer Vision and Pattern Recognition, New Orleans, LA, USA, 18–24 June 2022; pp. 10684–10695.
23. Ruiz, N.; Li, Y.; Jampani, V.; Pritch, Y.; Rubinstein, M.; Aberman, K. DreamBooth: Fine Tuning Text-to-Image Diffusion Models for Subject-Driven Generation. In Proceedings of the IEEE/CVF Conference on Computer Vision and Pattern Recognition, Vancouver, BC, Canada, 17–24 June 2023; pp. 22500–22510.
24. Hu, E.J.; Shen, Y.; Wallis, P.; Allen-Zhu, Z.; Li, Y.; Wang, S.; Wang, L.; Chen, W. LoRA: Low-Rank Adaptation of Large Language Models. In Proceedings of the ICLR 2022, Online, 25–29 April 2022 ; Volume 1, p. 3.
25. Esser, P.; Rombach, R.; Ommer, B. Taming Transformers for High-Resolution Image Synthesis. In Proceedings of the IEEE/CVF Conference on Computer Vision and Pattern Recognition, Nashville, TN, USA, 20–25 June 2021; pp. 12873–12883.
26. He, K.; Zhang, X.; Ren, S.; Sun, J. Deep Residual Learning for Image Recognition. In Proceedings of the IEEE Conference on Computer Vision and Pattern Recognition, Las Vegas, NV, USA, 27–30 June 2016; pp. 770–778.
27. Vaswani, A.; Shazeer, N.; Parmar, N.; Uszkoreit, J.; Jones, L.; Gomez, A.N.; Kaiser, Ł.; Polosukhin, I. Attention Is All You Need. *Adv. Neural Inf. Process. Syst.* **2017**, *30*, 5998–6008.
28. Dosovitskiy, A.; Beyer, L.; Kolesnikov, A.; Weissenborn, D.; Zhai, X.; Unterthiner, T.; Dehghani, M.; Minderer, M.; Heigold, G.; Gelly, S.; et al. An Image Is Worth 16×16 Words: Transformers for Image Recognition at Scale. *arXiv* **2020**, arXiv:2010.11929.
29. Liu, Z.; Lin, Y.; Cao, Y.; Hu, H.; Wei, Y.; Zhang, Z.; Lin, S.; Guo, B. Swin Transformer: Hierarchical Vision Transformer Using Shifted Windows. In Proceedings of the IEEE/CVF International Conference on Computer Vision, Montreal, BC, Canada, 11–17 October 2021; pp. 10012–10022.
30. Mehta, S.; Rastegari, M. MobileViT: Light-Weight, General-Purpose, and Mobile-Friendly Vision Transformer. *arXiv* **2021**, arXiv:2110.02178.
31. Hinton, G.; Vinyals, O.; Dean, J. Distilling the Knowledge in a Neural Network. *arXiv* **2015**, arXiv:1503.02531.
32. Paszke, A. PyTorch: An Imperative Style, High-Performance Deep Learning Library. *arXiv* **2019**, arXiv:1912.01703.
33. Brown, T.; Mann, B.; Ryder, N.; Subbiah, M.; Kaplan, J.D.; Dhariwal, P.; Neelakantan, A.; Shyam, P.; Sastry, G.; Askell, A.; et al. Language Models Are Few-Shot Learners. *Adv. Neural Inf. Process. Syst.* **2020**, *33*, 1877–1901.

34. Habib, G.; Kaleem, S.M.; Rouf, T.; Lall, B. A Comprehensive Review of Knowledge Distillation in Computer Vision. *arXiv* **2024**, arXiv:2404.00936.
35. Naddaf Shargh, A. Deep Learning Methods for Defect Analysis in Industrial Ultrasonic Images. Doctor of Engineering Thesis, Lamar University, Beaumont, TX, USA, 2025; Available online: `https://www.proquest.com/openview/2f717dbb11a339f3c9747c35937880cc/1?pq-origsite=gscholar&cbl=18750&diss=y` (accessed on 31 October 2025).

Index

scattered porosity, 53–55, 68, 74, 87–88

For Product Safety Concerns and Information please contact our EU representative GPSR@taylorandfrancis.com Taylor & Francis Verlag GmbH, Kaufingerstraße 24, 80331 München, Germany

Batch number: 10823959

Printed by Printforce, the Netherlands